Contents

WITHDRAWN FROM LIBRARY

Medicine at the Border

Also by Alison Bashford

IMPERIAL HYGIENE: a Critical History of Colonialism, Nationalism and Public Health

PURITY AND POLLUTION: Gender, Embodiment and Victorian Medicine

ISOLATION: Places and Practices of Exclusion (*co-edited with Carolyn Strange*)

CONTAGION: Historical and Cultural Studies (*co-edited with Claire Hooker*)

Medicine at the Border

Disease, Globalization and Security, 1850 to the Present

Edited by Alison Bashford

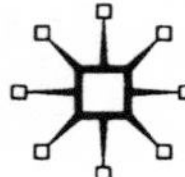

First published 2006 by
PALGRAVE MACMILLAN
Houndmills, Basingstoke, Hampshire RG21 6XS and
175 Fifth Avenue, New York, N. Y. 10010
Companies and representatives throughout the world

PALGRAVE MACMILLAN is the global academic imprint of the Palgrave Macmillan division of St. Martin's Press, LLC and of Palgrave Macmillan Ltd. Macmillan® is a registered trademark in the United States, United Kingdom and other countries. Palgrave is a registered trademark in the European Union and other countries.

ISBN-13: 978–0–230–50706–7 hardback
ISBN-10: 0–230–50706–9 hardback

This book is printed on paper suitable for recycling and made from fully managed and sustained forest sources.

A catalogue record for this book is available from the British Library.

Library of Congress Cataloging-in-Publication Data
Medicine at the border : disease, globalization and security, 1850 to the present / edited by Alison Bashford.
p. cm.
Includes bibliographical references and index.
ISBN 0–230–50706–9
1. Globalization–Health aspects. 2. Public health–Political aspects.
3. World health. I. Bashford, Alison, 1963–

RA441.M44 2006
362.1–dc22 2006049408

10 9 8 7 6 5 4 3 2 1
15 14 13 12 11 10 09 08 07 06

Transferred to Digital Printing 2008

List of Tables and Figures

Acknowledgments

The inspiration for this collection was a meeting at the University of Sydney in July 2004, 'Medicine at the Border'. My thanks to all participants and assistants at that productive conference, to Stephen Garton, Dean of the Faculty of Arts, for his support, and to the University for granting me a budget from the International Development Fund. Jill Gordon and Joanne Finkelstein have been fabulous colleagues over these particular years, in developing a Medical Humanities program. The book has also been assisted by an award from the British Academy, the Australian Academy of the Humanities and the Academy of the Social Sciences in Australia.

The Wellcome Trust Centre for the History of Medicine, UCL, provided office accommodation, resources and collegiality over the summer of 2005, where much of the editorial work was done. My thanks to the Wellcome Trust, and to Hal Cook, Anne Hardy and Sally Bragg at the Centre. The book has benefited from the suggestions of Joseph Melling, Anne Borsay, Carolyn Strange, Hans Pols, and several anonymous reviewers. I am grateful for their time and expertise. The process has been ably assisted by a number of researchers at my home department, the School of Philosophical and Historical Inquiry, University of Sydney, and my thanks to them all: Émilie Paquin, Catie Gilchrist, Jennifer Germon, Bernadette Power. Thanks also to Oscar Brunton and Tessa Brunton.

Brown, Cueto and Fee's chapter appears in a shorter version in *American Journal of Public Health* (January 2006): 62–72. It is reprinted here with permission from the American Public Health Association.

List of Abbreviations

BSE	Bovine Spongiform Encephalopathy
BWC	Biological Weapons Convention
CDC	Centers for Disease Control and Prevention
CEPR	Centre for Emergency Preparedness and Response
CsCDC	Consultants in Communicable Disease Control
CSR	Communicable Diseases Surveillance and Control
CMH	Commission on Macroeconomics and Health
DFID	Department for International Development
EGALE	Equality of Gays And Lesbians Everywhere
EID	Emerging Infectious Diseases
EU	European Union
FCO	Foreign and Commonwealth Office
GPHIN	Global Public Health Intelligence Network
GOARN	Global Outbreak Alert and Response Network
HIV/AIDS	Human Immuno-deficiency Virus/Acquired Immune Deficiency Syndrome
ICC	Isthmian Canal Commission
ICRC	International Committee for the Red Cross
LN	League of Nations
MAE	(French) Ministry of Foreign Affairs
MDG	Millennium Development Goals
MDM	Médecins du Monde
MSF	Médecins Sans Frontières
MOH	Medical Officer of Health
NIH	National Institutes of Health
NHS	National Health Service
NGOs	Non-Governmental Organization
ODA	Overseas Development Aid
Oxfam	Oxford Famine Relief Committee
PEPFAR	President's Emergency Plan for AIDS Relief
RF	Rockefeller Foundation
SARS	Severe Acute Respiratory Syndrome
TOPOFF	Top Officials
UK	United Kingdom
UNCTAD	United Nations Conference on Trade and Development
UNFPA	United Nations Population Fund
UNRRA	United Nations Relief and Rehabilitation Administration
US	United States (of America)
WHO	World Health Organization
YFC	Yellow Fever Commission

Notes on Contributors

Alison Bashford is a medical historian based at the University of Sydney. She has published on the history of public health and colonialism in the modern period, and on gender and medicine. Books include *Imperial Hygiene* (2004) *Purity and Pollution: Gender, Embodiment and Victorian Medicine* (1998) and *Isolation: Places and Practices of Exclusion* co-edited with Carolyn Strange (2003). Her current projects are an intellectual history of the problem of world population in the first half of the twentieth century, an international history of eugenics, and a biography of the geographer Thomas Griffith Taylor.

Sanjoy Bhattacharya is lecturer at the Wellcome Trust Centre for the History of Medicine at University College London, and his work deals with preventive and curative medicine in nineteenth and twentieth-century South Asia, as well as different aspects of international health history. He is author of *Propaganda and Information in Eastern India, 1939–45: A Necessary Weapon of War* (2001), *Expunging Variola: Public Health and the Control and Eradication of Smallpox in India, 1850–1977* (with Mark Harrison and Michael Worboys, 2005), and *Expunging Variola: The Control and Eradication of Smallpox in India, 1947–1977* (2006). His edited book, *Imperialism, Medicine and South Asia: A Socio-political Perspective, 1800–1950*, is also forthcoming, and he continues to edit *Wellcome History* (a post he was appointed to in 2002) and the New Perspectives in South Asian History Series (brought out by Orient Longman India Ltd and Sangam Books UK).

Theodore M. Brown is Professor and Chair of History and Professor of Community and Preventive Medicine and of Medical Humanities at the University of Rochester in Rochester, New York (USA). He earned his Ph.D in history of science at Princeton University and studied as a post-doctoral fellow at the Johns Hopkins University Institute of the History of Medicine. He has published numerous articles and book chapters on the history of medicine, public health and health policy and has authored *The Mechanical Philosophy and the 'Animal Oeconomy': A Study in the Development of English Physiology During the Seventeenth and Eighteenth Centuries*. In 1997 he and Elizabeth Fee published their co-edited and substantially co-authored *Making Medical History: The Life and Times of Henry E. Sigerist*. He is Contributing Editor for History with the *American Journal of Public Health* and is Editor of the book series, *Rochester Studies in Medical History* at the University of Rochester Press.

Richard Coker is Reader in Public Health and Policy at The London School of Hygiene and Tropical Medicine. For 3 years, from 1995, he headed the tuberculosis service at St. Mary's Hospital, London. In 1997 he was granted a Harkness Fellowship by the Commonwealth Fund of New York and spent a year as visiting scholar to the School of Public Health at Columbia University, New York, where he researched the public health responses to the contemporary New York tuberculosis epidemic. His book, *From Chaos to Coercion: Detention and the Control of Tuberculosis* resulted from this work and was published in 2000. On his return to Europe in 1999 he led the Task Force on the Urgent response to STIs in eastern Europe for the World Health Organization, and later worked at The Wellcome Trust Institute as a Research Associate researching domestic legislative responses to infectious diseases. His current research interests include communicable disease control in eastern Europe, and legislative responses to public health threats, and international planning and policy responses to pandemic influenza.

Ian Convery is Lecturer in Human Geography at the University of Central Lancashire. His doctoral research at Lancaster University was in the community management of natural resources, in particular the relationship between forest peoples and National Park Authorities in Central Mozambique. He has since worked mostly around community development, focusing on issues of health and migration, and has published widely in these areas. Recent articles and chapters include emotional geographies of the 2001 UK Foot and Mouth Disease epidemic; 'citizen' versus 'professional' epidemiologies; and South Asian migrants in the UK textile industry.

Marcos Cueto is a Peruvian historian and a professor at the Universidad Peruana Cayetano Heredia of Lima. He received his Ph.D from Columbia University. He is the author of books and articles on the history of medicine and public health in Latin America. He has received grants from the Guggenheim, Fulbright and the Rockefeller Foundations. In 2003–4 he co-chaired (with Elizabeth Fee) the History Working Group of the Rockefeller Foundation-sponsored program 'Joint Learning Initiative Human Resources for Health and Development'. His books include *The Return of Epidemics, Health and Society in Peru of the 20th Century* (2001) and *El Valor de la Salud, una historia de la Organizacion Panamericana de la Salud* (2004).

Elizabeth Fee is Chief of the History of Medicine Section, the National Library of Medicine, Bethesda. Previously she was Professor of History and Health Policy at the School of Hygiene and Public Health, the Johns Hopkins University. With Theodore Brown, she is author of *American Public Health Association. Conflict and Controversy: from Medical Care Policy to the Politics of Environmental Health* (1998) and co-editor of *Making Medical History: The Life and Times of Henry E. Sigerist* (1997). She is also co-editor of

Women's Health, Politics, and Power: Essays on Sex/Gender, Medicine, and Public Health (1994) and *AIDS: The Burdens of History* (1988).

David P. Fidler is Professor of Law and Harry T. Ice Faculty Fellow at Indiana University School of Law, Bloomington, USA. He is one of the world's leading experts on international law and public health, with an emphasis on infectious diseases. He has published widely in both law journals and public health periodicals. His books include *SARS, Governance, and the Globalization of Disease* (2004), *International Law and Public Health* (2000), and *International Law and Infectious Diseases* (1999). He has served as an international legal consultant to the World Health Organization (on international trade law and the International Health Regulations), the US Department of Defense's Defense Science Board (on defense against biological terrorism), and various non-governmental organizations (on global health and non-proliferation issues). He is a Senior Scholar at the Center for Law and the Public's Health at Georgetown and Johns Hopkins Universities, and he has been a Fulbright New Century Scholar. He has degrees from the University of Kansas, Harvard Law School, and the University of Oxford.

Claire Hooker is a historian of science, medicine and health. She currently holds the NH&MRC Sidney Sax Post-doctoral Fellowship in Public Health and is based at the University of Toronto investigating the cause of and social response to health scare situations. She has previously worked with prominent public health advocate Professor Simon Chapman on the political history of tobacco control. She has co-edited *Contagion: Historical and Cultural Studies* (2001) with Alison Bashford, and her book *Irresistible Forces: Australian Women in Science* appeared in 2004.

Alan Ingram is Lecturer in Geography at University College London, working on globalization, geopolitics and security with particular reference to questions of health and disease, and is the author of a number of forthcoming articles on these themes. He previously managed the UK Global Health Programme on Health, Foreign Policy and Security at the Nuffield Trust, and has taught at the University of Cambridge and the London School of Economics.

Renisa Mawani is an Assistant Professor in the Department of Anthropology and Sociology at the University of British Columbia. She has published in the areas of moral regulation; (post)colonialism, law, and space; histories of Chinese migration to British Columbia; and on the legal constructions of Aboriginal identity. Her publications have appeared in journals including *Law/Text/Culture*, *Social and Legal Studies*, and *Canadian Journal of Law and Society*. She is currently working on two books. The first is a socio-legal history of law and colonialism in late nineteenth and early twentieth century British Columbia, and the second is a (post)colonial legal history of Vancouver's Stanley Park.

Eric Mykhalovskiy is an Assistant Professor in the Department of Sociology, York University. His main interest is in the social organization of health knowledges. He has pursued this interest in research on evidence-based medicine and on HIV/AIDS that has been published in such journals as *Social Science and Medicine, Health,* and *Social Theory and Health.*

Alexandra Minna Stern is Associate Director of the Center for the History of Medicine and Associate Professor of Obstetrics and Gynecology and American Culture at the University of Michigan. Her book *Eugenic Nation: Faults and Frontiers of Better Breeding in Modern America* was published in 2005 by University of California Press. Dr Stern recently received a National Endowment for the Humanities Fellowship and National Library of Medicine Publication Grant to support her next research project on the history of genetic counseling in the United States.

Carolyn Strange is Director of Graduate Studies, Centre for Cross-Cultural Research, the Australian National University. Carolyn Strange's most recent books are *True Crime, True North: The Golden Age of Canadian Pulp Magazines* (with Tina Loo) and *Isolation: Places and Practices of Exclusion* (co-edited with Alison Bashford). She has published widely on the interface between tourism, tragedy, and troubling pasts. Recent articles focus on the phenomenon of prison history tourism, comparing Port Arthur, Australia; Robben Island, South Africa; and Alcatraz, USA. She extended this work in a study of Kalaupapa, a National Historic Park in Hawaii, which was formerly a colony of forced exile for persons with leprosy. Currently adjunct Professor of History and Criminology, University of Toronto, she was a resident of Toronto during the SARS crisis of 2003.

Miriam Ticktin is Assistant Professor in Women's Studies and Anthropology at the University of Michigan. She received a Ph.D in Cultural Anthropology from Stanford University in 2002, and a Ph.D in Medical Anthropology in 'co-tutelle' with the École des Hautes Études en Sciences Sociales (EHESS) in France. She is currently working on her book, *Between Justice and Compassion: The Politics of Immigration and Humanitarianism in France,* about the fight for social justice of undocumented immigrants, as well as co-editing a volume called *Government and Humanity* which explores new conceptions of the category humanity. She has published on issues related to the politics and ethics of medical humanitarianism, immigration and asylum-claimants in Europe, the relationship of sexual violence to the politics of immigration, and the role suffering plays in legal claims. Her new research explores the relationship between biology, citizenship and global capitalism.

Lorna Weir is Associate Professor in the Department of Sociology, York University, Canada. She has published widely on the government of human reproduction, including *Pregnancy, Risk and Biopolitics: On the Threshold of the*

Living Subject (2006). Her current research is focused on two areas: news and security in global public health surveillance (with Eric Mykhalovskiy) and the biopolitics of advanced modernity.

John Welshman was educated at the Universities of York and Oxford and is Senior Lecturer in Public Health at the Institute for Health Research at Lancaster University, UK. He is the author of *Municipal Medicine: Public Health in Twentieth Century Britain* (2000); *Underclass: A History of the Excluded, 1880–2000* (2006); and editor, with Jan Walmsley, of *Community Care in Perspective: Care, Control, and Citizenship* (2006). He is currently working on *Policy, Poverty, and Parenting: From Transmitted Deprivation to the Cycle of Disadvantage* (forthcoming, 2007).

Patrick Zylberman is senior researcher at CERMES (Centre de Recherche Médecine, Sciences, Santé et Société), CNRS-INSERM-EHESS, Paris, France. He jointly authored with Lion Murard *l'Hygiène dans la République. La santé publique en France, ou l'utopie contrariée, 1870–1918* (1996). His research focus is on public health and state building in twentieth-century Europe, and on public health in France from the 1920s to the 1940s. Among his recent publications is: 'Fewer Parallels than Antjtheses: René Sand and Andrija Stampar on Social Medicine, 1919–1955', *Social History of Medicine*, 2004.

1
'The Age of Universal Contagion': History, Disease and Globalization

Alison Bashford

Medicine at the Border explores the pressing issues of border control and infectious disease in the nineteenth, twentieth and twenty-first centuries, in the 'age of universal contagion'.[1] This book places world health in world history, microbes and their management in globalization, and disease in the history of international relations, bringing together leading scholars on the history and politics of global health. Together, the authors show how infectious disease has been central to the political, legal and commercial history of nationalism, colonialism, and internationalism, as well as to the twentieth-century invention of a newly imagined space for regulation called 'the world'.

This is a modern history of a world with markedly more health, and less acute infectious morbidity and mortality than previous centuries, in some places. It is also the history of a divided world, where the manifestly unequal distribution of the benefits of modern medicine and public health marks the division between North and South, West and East possibly more starkly than any other factor. Thus, analysis of global health raises a history of medicine, but it also raises a history of geopolitics. The geopolitical aspect not only concerns the historical geography of disease itself, but also that of disease *management:* the reduction and prevention of microbes and illness is rarely the only outcome of disease control, even if it is an important one. The chapters collected here squarely address the under-recognized place of disease control in the history of national and international governance, and in the processes of globalization over the modern period. *Medicine at the Border* goes some way towards developing what might be called a world history of the geopolitics of disease prevention.

Infectious disease management often has spatial implications, and uses spatial measures of prevention, reduction and eradication: this is as true in the twenty-first century as it was in the nineteenth century. For this reason, borders of many kinds, and in many places, so often recur in practice: quarantine lines and migration screening, once a stethoscope to the bare chest, now fever scanning at airports; trade barriers against BSE; home isolation for the suspected carrier of the SARS virus; targeted vaccination in social and spatial circles around a remaining smallpox victim. There is a geopolitics to each of these. But to focus on what happens *at* these borders, is to miss another scale of geopolitics and disease management altogether. For the politics of disease control concerns the governance of *this side* and crucially *that side*, of the border as well. As we shall see, infectious disease control – and the relief and prevention of suffering – has not infrequently been a rationale for all kinds of formal and informal intervention beyond a local jurisdiction, beyond a sovereign state. Over and over again, the aspiration to promote health and prevent disease has resulted in pre-emptive activity beyond the border: European powers sought to intervene in Ottoman rule to prevent the spread of cholera; the US military, followed by the Rockefeller Foundation, embarked on major yellow fever eradication campaigns in Cuba, Panama, Puerto Rico; quarantine lines in Africa offered a clear and politically useful demarcation for new 'international' borders between Sudan and Egypt, between Uganda, French Congo and Belgian Congo.[2] This kind of geopolitics of disease prevention continues in the realm of health aid and development:[3] for example, assistance with HIV/AIDS treatment will be given *there* (in the form of aid) but not *here* (in the form of state-funded treatment of refugees). And other kinds of aid workers cross national and political borders defiantly, armed with ideas of 'humanitarianism' and 'universalism' along with vaccines and antibiotics: they aim for, and try to practice within, a world 'sans frontières'.

While once disease prevention and geopolitics were simply related, more recently the former has become a vehicle for, and even an instrument of, the latter. The intense twenty-first century manifestations of defensive nationalism, disease and security on the one hand, and global flow, supranational surveillance technologies, actual and imminent world pandemics on the other, suggest a need to think about the provenance of these connections, their effects in the past, and to temper assessments of their alleged novelty, while at the same time recognizing a world linked in time and space in ways altogether new.

Part I of this book, and of this introduction, deals with the connections between national histories and the emergence of international and world health structures over the twentieth century. Authors interrogate the internationalization of world systems of epidemic management, of eradication dreams, of colonization, decolonization and world health governance. The impossibility (and therefore historical inaccuracy) of separating out colonial and national histories in the genealogy of world health becomes apparent. Part II discusses the issue of territorial health regulation, of medico-legal border control and movement of people across national lines, both historically and in the twenty-first century. In Part III, authors pursue aspects of late twentieth and twenty-first century global disease and security. They examine contemporary global epidemic surveillance and information networks, flows of information, microbes and fear, which increasingly bypass the nation-state, but cannot do so altogether.

World health: national and colonial histories

Several historical lines merged to create 'world health' as a problem, a project and a possibility in the twentieth century: colonial medicine, national territorial defense imperatives, international convention and agreement on both trade and disease, regionally-interested organizations. The domain of health and disease regulation was by no means incidental to the consolidation of nineteenth and twentieth-century territorial nation-states and to the related phenomenon of international relations. Historians have rightly detailed the quarantine and so-called 'sanitary' conferences of the mid to late nineteenth century, as the direct precursors to the early twentieth-century international organizations of health,[4] and to the evolution of Westphalian systems in practice.[5] The reverse has also been argued: that a new internationalism created national public health measures. 'It is only against the background of medical internationalism', write Stern and Markel, 'that we can begin to understand the elaboration of the United States Public Health Service regulations on immigrants inspection, quarantine, and vaccination in the early 20th century'.[6]

There was a string of international meetings from 1851, initially taking up the question of cholera, and later plague and yellow fever. As Brown, Cueto and Fee summarize in their chapter, in 1902 an International Sanitary Office of the American Republics was established which became the Pan American Sanitary Bureau. In Europe, the International Sanitary Conference discussed the need for a permanent international body and the Office international d'hygiène publique resulted, based in Paris (1907). After World War I, the League of Nations created an Epidemic Commission to deal with typhus fever in Eastern Europe,[7] and in 1923 the Health Organization of the League of Nations was established with four areas of work: epidemiology; technical studies; study tours; and the 'intelligence' work of the Far Eastern Bureau at Singapore.[8] Alongside the proliferation of formal intergovernmental organizations for health entered philanthropic organizations like the various renditions of the Red Cross, and private US philanthropic organizations, most significantly the Rockefeller Foundation with its International Health Board established in 1913.[9] The World Health Organization succeeded both the League of Nations Health Organization and Rockefeller's International Health Board after World War II.

Zylberman shows the centrality of cholera and the Mecca pilgrimage in shaping the earliest international discussions and agreements over infectious disease, and in establishing precedents for European powers' sanitary intervention into Ottoman-ruled territories and peoples. Zylberman analyzes the complicated relations between European powers and the Ottoman

Empire, where the latter's apparent failure to contain disease and to implement preventive measures was a risk to Europeans. This betrayed it as a 'weak state' and animated European public health officials to act. Zylberman argues that this was 'pre-emption' based on the need for (health) security of Europeans. On the one hand, European powers were increasingly subscribing to Westphalian principles of non-intervention. On the other, they were in practice intervening through a range of colonial and occupying structures, and through a discourse of 'civilization' whereby another state's incapacity to be 'civilized' in a 'sanitary' sense was a justification for intervention.

The nineteenth-century sanitary conventions and the early twentieth-century organizations are most often discussed as the predecessors of later twentieth-century world health. But there is another line of development: colonial medicine and tropical medicine. The study of the 'diseases of warm climates' institutionalized into the discipline of tropical medicine in the late nineteenth century at sites like the London School of Hygiene and Tropical Medicine, the Pasteur Institutes in Paris and the French colonies, and the Johns Hopkins Medical School. As many scholars have shown, tropical medicine was institutionally, politically and intellectually about the large and broad project of colonization,[10] it was always implicated in the (medicalized) question of geography and place. Originally concerned with the health of Europeans and Anglo-Americans 'elsewhere', that is, in colonial situations and in 'the tropics', the discipline gradually developed research and clinical interests in indigenous people, locals, those who were understood to belong to place by virtue of history, and of race, climate, geography and constitution.[11]

This historical scholarship has focused largely on British and French tropical medicine, yet as Alexandra Minna Stern points out, the US history of tropical medicine and colonial medicine merits further attention.[12] Stern examines here the extraordinary success (at one level) of the US military-sanitary campaign in Cuba, which in one year reduced yellow fever morbidity to zero. As a result of that success a similar strategy was implemented in the Panama region, during the building of the Canal between 1904 and 1914. Thereafter, the Rockefeller Foundation assumed real interest in and much control over disease eradication programs across Latin America. The connections in this story between commerce, international relations, US military colonialism, and a philanthropic public health were tight indeed. Stern shows at once an important US axis on which colonial medicine turned, a clear argument for the racial systems which underpinned and were perpetuated by these health campaigns, and a specific example of how tropical medicine was one of the roots of international health.

The history of world health cannot be understood as anything but merged formations of colonial, national, and 'world' politics, played out on specific local ground. Stern's case study shows how these public health campaigns were often simultaneously about the nation (US security and commerce) about colonization and colonized people and places (Cuba, Panama) and about emerging meanings of 'international' (Rockefeller's public health interventions across Latin America). Indeed the Rockefeller Foundation, through its extraordinary level of funding and influence was a hinging factor between colonial medicine and international health, shaping, for example, *both* 'colonial' institutions like the London School of Hygiene and Tropical Medicine *and* 'international' institutions like the League's Far Eastern Bureau in Singapore, and indeed the League of Nations itself.[13] Sanjoy Bhattacharya's study of another eradication campaign, the famous instance of the eradication of smallpox in India, shows again the complicated interplay between the local, the national and the international. Challenging received stories of the victorious prominence of WHO personnel, Bhattacharya details the tense and complicated politics and pragmatics of local implementation and of the Indian government's investment in eradication. Rather like Patrick Zylberman's analysis of the problem of 'intervention' on the part of the European powers into the Ottoman Empire, the smallpox eradication campaign was one which constantly encountered tensions between the international body and the sovereign authority and interests of the nation.

That Stern writes of the Panama Canal and Zylberman of the Suez Canal suggests that national borders are not the only territorial demarcations at work in this history of global health and disease control. This is not only about nations, but also about the formation of geopolitical 'regions'. These were often regions of colonial influence, often less formally demarcated than nations, but not exclusively so: Coker and Ingram's discussion of Europe as a contemporary health region is a case in point. Alongside the development of 'international' institutions like the Paris Office or the League of Nations Health Organization were expressly regional institutions, such as the Pan American Health Organization as well as the League's own Far Eastern Bureau based in Singapore.[14]

A shifting lexicon is significant here. Brown, Cueto and Fee explore the changes from 'world health' to 'international health' to 'global health', focusing on the fortunes of WHO since 1948. Weaving together the politics and pragmatics of the Organization and its leaders, as well as changing understandings of 'health' from, for example, disease eradication to primary health care, they show the use of the discourse of 'global' for a renewal of WHO as an organization. Brown, Cueto and Fee's chapter contributes to what is currently a fairly thin historical scholarship on WHO, in contrast to the rich historical work on the League's organizations, and on the International Health Board.[15] Their work opens up further questions about 'the international' to be historically (and geographically) scrutinized. For example, at the beginning of the twentieth century, 'international' often signified Europe: the so-called 'international' health organizations (at least the early examples like the Paris Office and the League of Nations Health Organization) were at core, in orientation, and in interest, regional European organizations. This is not to diminish their significance, and certainly not suggested in ignorance of ventures like the Far Eastern Bureau, the influence of non-European personnel, and the intermittent inclusion of the Americas. Rather it is to suggest that the primary drive of the Health Organization, like the League of Nations itself, was the reorganization of Europe, and both internal and external European security (including colonial possessions and mandated territories). The point is that European-based organizations and meetings in this period could apply the label 'international' in a sustained way, whereas American regional efforts were geographically marked as 'Pan American'. It is salient that, as Brown, Cueto and Fee show, the 1902 International Sanitary Office of the American Republics became after World War I the Pan American Sanitary Bureau and then the Pan American Health Organization. Likewise, Pacific-rim organizations were marked geographically as 'Pan Pacific' (for example the Pan Pacific Science Congresses). There is an opportunity here to apply historian Dipesh Chakrabarty's call to 'provincialize Europe' in our discussion of European 'international' health organizations, and to think about the history of regionalization, as well as colonialism and nationalism, in the development of a twentieth-century global health.[16]

If 'international' at the beginning of the twentieth century stood broadly for 'Europe', by the decades of decolonization after World War II, 'international' in the domain of health came to mean the health of 'developing' countries: largely infectious disease prevention and eradication programs in the so-called third world.[17] In a strange turn, it was at this point that 'international' health fully inherited tropical and colonial medicine; or to put it another way, tropical medicine itself was decolonized. And yet numerous scholars – not least Weir and Mykhalovskiy in their chapter on geopolitics and public health surveillance – argue that later twentieth-century world health administration has inherited this history of health and colonization, and retains a strongly neo-colonial character. Obijifor Aginam suggests that this colonial inheritance has created a world health culture oscillating between 'global neighbourhood' and 'universal otherhood'.[18]

National security: territory, migration and border regulation

The geopolitics of disease prevention has often operated through, and linked, nationalism and the policing of sovereign territory. From the early nineteenth century both maritime and land borders became closely regulated places for the inspection of the goods of commercial exchange, as well as vessels and animals. This is why, in many modernizing bureaucracies, quarantine officials were typically located within the broader government office and power of 'customs'. But nineteenth-century quarantine law typically governed the movement and traffic of goods, animals *and* humans. With the emergence of European nation-states and their colonial extensions over the nineteenth century, and with increasingly bureaucratized administrative government, disease was checked by border inspections of people – their bodies, their identity and their documents. The documents of health, of being disease-free (or more likely coming from a disease-free town or region) existed as a system prior to the widespread use of identity documents (the passport or the visa, for example). Thus one of the factors which made jurisdictional (increasingly meaning national) borders meaningful was the checking of health documentation and of people's bodies for signs of infectious disease, and indeed, for signs of disease prophylaxis – vaccination. These procedures made borders more than abstract lines on maps, but a set of practices on the ground.

On a world basis (but, as we shall see with some most interesting exceptions), immigration law and public health law became connected in the late nineteenth and early twentieth centuries. This was a regulatory response to the phenomenon of mass movement, of circulating diasporic labor, of migrants, pilgrims, and refugees. A considerable body of scholarship details the health clauses – the 'loathsome disease' clauses as they were often called – in immigration law of various jurisdictions.[19] In Europe, cholera and the Mecca pilgrimage constituted the 'eastern question' as a 'health question'.[20] In the US, Russian-Jewish migration embedded quarantine screening in entry procedures,[21] while on the west coast and elsewhere in the nineteenth-century Pacific, Chinese indentured laborers and gold-seekers were singled out for regulation. Chinese Exclusion Acts emerged from the 1880s onwards, cementing ideas about, and joint regulation of, race, disease, territory and nationalism. Ironically, in the 1950s and 60s, decolonizing nations throughout South East Asia borrowed migration law from the 'colonial-settler' nations, and wrote similar health clauses into their new national statutes.[22] Through the implementation of these powers, national populations were literally shaped, territories were marked, and inclusions and exclusions on all kinds of bodily criteria were implemented.

But policing national territory was rarely about complete exclusion, historically or currently. Rather, in both quarantine and immigration domains, it was usually about monitoring entry and selectively including. Recently historians have shown the close connections between border screening processes and precarious inclusion into territory and civic identity: a rich historiography of health and citizenship is emerging. Both Fairchild and Shah have demonstrated this on either side of the US continent. Decades of migrant health screening at Ellis Island, New York, did not rest on any sort of legitimate microbiological or epidemiological rationale, Fairchild argues. Rather, it performed a more complex function of initiation into an industrial culture.[23] On the west coast, Shah has shown how initial exclusion of Chinese in the late nineteenth century, became a provisional incorporation of Chinese communities into the US civic body by the 1920s and 30s, but one dependent on 'standardizing Chinese conduct and living spaces according to American hygienic norms'.[24]

In this book, several extensions to this scholarship are offered, as well as new case studies which detail interesting exceptions to the 'exclusion' model. In a study of health screening in the United Kingdom and Australia, Convery, Welshman and Bashford compare the quintessential instance of national medico-legal border control (Australia) and, as it turns out, one of the most exceptional instances (the UK). Elsewhere, Bashford has elaborated the deep significance of quarantine and health clauses of immigration restriction acts to the demarcation and defense of Australian territory (as an island-continent), to Australian nationalism and specifically to the White Australia Policy. These had a very particular manifestation and connection to nationalism in Australia, but, as she has argued, the conflation of health and immigration border screening with the element of racial exclusion was 'rather more ordinary than extraordinary' for the period (of the early to mid-twentieth century).[25] Until recently, the United Kingdom's history was starkly different. Certainly in the UK, there has been a long popular (and sometimes expert) linking of contagion, race, disease and dreams of exclusion and cultural/racial homogeneity, but this was not rendered into law and official policy in the way that was so common elsewhere (not only Australia, but the US, Canada, New Zealand for example). There has been minimal formal linking of health and immigration powers in the UK. This highlights that in the world history of 'medicine at the border', it was the colony-nations of settlement and importantly the destinations of the Chinese diaspora in the nineteenth century which created the legacies of joint health and immigration law and regulation. This history of UK 'exceptionalism' is an interesting reversal of the more usual center-periphery dynamic of colonial/imperial history. But as these authors show, and as Coker and Ingram demonstrate in their chapter on more recent developments, the future may look very different indeed. In the last few years, the UK government and opposition have sought to bring the UK 'into line' as it were: they are flagging for implementation what is known as 'the Australian model' of rigid pre-entry screening for various infectious diseases.

A different exception to the dominant history of the exclusionary capacity of border screening and territorial nationalism, is the case of France. In an important study, Miriam Ticktin details a counter discourse to exclusion in the French political and philosophical tradition of universalism and humanitarianism, expressed in the history of 'medical humanitarianism'. She details the history of Médecins Sans Frontières (MSF) within this tradition, and the reach of the idea of medical humanitarianism (as well as MSF personnel) into French government and law. Partly because of the character of French colonialism which incorporated colonial land and people into

France itself, and partly because of the philosophy and idea of 'the universal' (and therefore the diminution of territorial, sovereign or civic human difference), the French government recognizes the right of sick people to make a claim to be treated within France. If other national histories are dominated by the idea that illness, defect, and disease render people ineligible for, or unlikely to receive, entry, in France the reverse is the case. And if in so many other contexts historically, a 'public charge' argument has been written into immigration law (that is, if the person is likely to become a cost to the community through health and welfare dependence they may be refused entry or deported) the French case certainly represents the reverse principle. But as Ticktin shows, there is a subsidiary history of colonial relations also at work in French medical humanitarianism. Rather like Shah's precariously included Chinese-US citizens,[26] people residing in France on grounds of their illness must remain ill, in order to literally stay: despite a rhetoric of universalism, they must remain in a position of dependency in, and on, the French state, never quite equal, never quite citizens.

While Convery, Welshman and Bashford compare the Australian and UK systems of health screening, it is also worth thinking about the French and British instances comparatively. In the French case, there is legal principle for positive action: the law provides for entry on the grounds of illness, on the principle of universal medical humanitarianism. In the UK, there is no such pro-active law or regulation, but much the same thing happens in practice. That is (at least for the moment) people diagnosed with, for instance, tuberculosis on entry are not excluded and made to undertake treatment in their country of origin at their own cost (as in the Australian case), but may indeed enter the UK, and will be followed up at their local destination, by their local health service. What adds a further layer to these histories of illness, exclusion and inclusion is the current phenomenon of so-called 'health tourism', wherein some nations are more open than others, to this particular kind of border crossing: certain *kinds of* (monied) sick people are temporarily admitted. Indeed scholarship on the current phenomenon, which is economic in nature, and turns on complicated and not necessarily one-way axes of North-South, East-West, could profit from being historicized in terms of nineteenth-century colonial 'health tourism'.[27]

Canada, like Australia, has strict screening policies and shares a not dissimilar history of conflated health and race exclusions, although one far less well known internationally. And yet both Canadian provincial and federal law imposed versions of immigration restriction acts that incorporated various kinds of health criteria, which were race-based in implementation and intention. Here, Renisa Mawani looks at a recent turn in this link between territorial policing and medical policing in Canada – the 2002 Immigration and Refugee Protection Act and the (new) provision for the mandatory HIV testing of all immigrants. These measures, she argues, while determinedly race neutral, are nonetheless often race specific in effect. Border security, she shows, is problematically coming to be promoted as a primary preventive health measure, as if it were comparable to domestic distribution of health resources and preventive education campaigns.

Epidemiologist Richard Coker has argued against the efficacy of border screening for disease prevention and management.[28] Here, with geographer Alan Ingram, he explores disease and migration regulation in the European Union and in the UK. In a field which often prioritizes the spread of acute disease, in particular SARS and imminent bird flu, Coker and Ingram insist on the need to keep scholarly watch on the politics of management of chronic infectious diseases – HIV/AIDS, tuberculosis and malaria. Like Mawani, Coker and Ingram draw attention to dubious (in public health terms) distinctions drawn between medical humanitarianism as part of foreign policy and the avoidance of medical humanitarianism as part of domestic health policy. The former is politically expedient 'good' aid, while the latter is politically inexpedient health funding of 'foreigners' as asylum claimants or as intending migrants. There is, then, a strange disjunction between promoting health aid elsewhere and the increasing refusal to treat (for free) those people when they are in the UK. This is part of the long history of the geopolitics of disease management, about people being considered properly *in their place*, or improperly *out of place*.

Globalization: deterritorialized health?

Chapters in Part III of this book place these histories of global geographies of disease and disease management, of borders, nationalism, and internationalism, in the present. This section deals directly with new formations of security, international relations and public health, which are nonetheless derivative of this history. By the 1990s, as Brown, Cueto and Fee show, the nomenclature of 'global' health began reliably to replace 'international' health in the context of 'new and re-emerging diseases' and of economic globalization. This usage of 'global' has intensified after the experience of SARS, anthrax bioterrorism, and threats of avian flu. In other words, microbial threats became 'global' when they have impacted, actually or potentially, on the 'first world', and in particular on the US.

Scholarship on globalization typically draws a distinction between 'international' (where the administrative and legal unit of the nation is always present) and 'supranational', 'transnational' or 'global' (where organizations or systems do not rely on or refer to the nation as a basic defining unit). Globalization is, by one definition: 'the acceleration and intensification of mechanisms, processes, and activities that are allegedly promoting interdependence and perhaps, ultimately, global political and economic integration'.[29] Global economic forces, it is argued, undermine the independent capacities of the sovereign nation-state, and the territorial basis of the Westphalian system is threatened because 'social space is no longer mapped in terms of territorial places, territorial distances and territorial borders'.[30] Instant communication, daily global mass movement, financial cyber-transactions that are 'placeless', have, over the last decade created a radically new world that is 'supraterritorial' or, as is often claimed, 'deterritorialized'. Other commentators and scholars understand 'globalization' as economic and cultural 'westernization'.

Globalization, disease and its management are related in several ways. First, the transborder nature of microbes and disease, has been, without question, augmented with the frequency of travel. Second, there has been considerable use of supranational, fully global technologies and networks to track disease outbreak, as Weir and Mykhalovskiy discuss. Third, the deep investment in the 'development' idea of international health and world health, whereby the third world is developed in line with first world sanitary and health conditions as a way to secure disease-free regions, represents the 'westernization' dimension of globalization, for all its benefits in terms of morbidity and mortality. This latter aspect recalls, of course, Zylberman's argument on civilization and sanitary pre-emption.

For these reasons and more, historians of public health need to further enter scholarly discussion on globalization. They need to complete the third side of a scholarly triangle. On one side, there is a considerable literature on globalization, disease and health in the contemporary world.[31] On a second, there is increasing scholarly interest in thinking about globalization historically, a recent extension of both imperial historiography and world historiography.[32] But the connecting side of the triangle is underdeveloped: the historical study of disease and its management as part of the historical process of globalization. Again, while there is some historical discussion which picks up the idea of globalization and *disease* (especially in world historiography)[33] there is less on the idea of supranational disease *management*, 'supranational' public health, as it were.[34] The histories of 'international hygiene', 'international health', and 'world health' are certainly an aspect of a strictly international history (that is a history of international relations and 'internationalism'),[35] but they are also important sites to examine the emergence of ideas about the 'world' in world health, or 'the globe' in globalization.[36]

Globalization scholars often insist on the diminution of geography and territorial borders, but most authors in this collection argue for their ongoing significance. On the one hand, we learn from chapters in Part II of the long legacy of national border control for public health which the twenty-first century inherits. And we also see that, if anything, this historical legacy of territorial medico-legal border control is recently consolidating. On the other hand, however, the last two decades have indeed offered 'deterritorialized' methods of surveillance, methods not based on national territorial security, or on the (literal) ground of border surveillance. Weir and Mykhalovskiy study closely GPHIN, the Global Public Health Intelligence Network. Radically shifting from the 'world health' tradition and history of sharing nationally-secured epidemiological information, GPHIN has used global news/information/internet sources to pick up the possibility of disease outbreak: it deploys the supranational 'network' of the internet, rather than the geographical line of governmental border surveillance. But Weir and Mykhalovskiy conclude (as indeed did GPHIN personnel) that this information gained supranationally, was then ineffectual unless it could be verified by an authorized international body, the WHO. Originally bypassing the nation-state, and the conventional international order, GPHIN found that it could not do so altogether.

World health has been centrally about information flow and exchange, from its origins in the nineteenth-century sanitary conferences and in the knowledge-machines of imperial infrastructure. 'Epidemiological intelligence' as it was often called was, in many ways, the first imperative of the various early international organizations for health and for the prevention of infectious disease. Yet media and communication exchange is not only about the sharing of epidemiological knowledge between experts, but also the proffering of advice from experts to 'the people' about how a disease should be prevented or minimized.

Two chapters on SARS highlight the persistence of the national and local in a supposedly globalizing world, and the significance of perceptions of security and safety, risk and danger, generated by strategic and sometimes accidental coalitions of national and international agencies, local and global media. In a sharp cultural study of SARS in Toronto, Strange explores the intense public relations/international relations efforts to re-package the city as clean and safe. The representation of Toronto was a deeply commercial question, in that the city had traded for years on its reputation as both clean and safe, and found itself momentarily a dangerous place and a 'pariah state', as well as fully 'exoticized' in its sudden link with the 'epicentre', China and Hong Kong.[37] Strange shows how the historic linking of Chinese diasporic communities with disease returned to shape the cultural response to SARS in North America. But this story was not a straightforward repetition of past Chinese discriminations and exclusions, for Toronto had also long packaged and traded itself as a multicultural city, an ethnically 'diverse' city. And so, as Strange details, the city's PR managers – professional semioticians – found themselves with an odd and difficult representational problem. How to manage competing perceptions of safety and risk, the exotic and the secure, in this moment of intense global surveillance?

While migrants and refugees have historically been the problematized population in terms of global infectious disease, SARS problematized the tourist and the business traveler. One of the important facets of the SARS episode in 2003 was that it crystallized for those few months, and onto the everyday tourist and traveler, many of the spatial techniques of prevention and surveillance used more diffusely and permanently on migrants and refugees all over the globe. Claire Hooker details how the fascinating epidemic unfolded and offers an analysis of how authorities mobilized both 'old-fashioned' public health models (whereby the 'dangerous' were isolated in the quarantine tradition), and newer risk-based models (where 'at risk' groups were acted upon). On the one hand, authorities looked to the past and borrowed clumsily from old coercive quarantine controls. On the other, these prompted a future-looking 'preparedness planning' mentality, where health risk-minimization measures dovetailed with newly rigid national security measures. Hooker explores how health professionals have taken on the 'new normal' of constant bio-preparedness. Evident in each of these chapters is SARS as an epidemic of fear, in which the perception of security becomes as important as actual security.

The 2003 timing of SARS, as the follow-up to the 2001 attacks on the US and anthrax and smallpox scares in 2001/02, meant, of course, that terrorism, bioterrorism and epidemic disease became conflated. This occurred both intuitively at popular levels, but also deliberately at expert and institutional levels, especially but not only in the US. As Fider argues, public health and 'homeland' security are increasingly twinned, as geopolitical and geo-epidemiological issues. The newly intense bio-preparedness imperative threatens to dominate public health priorities, both nationally and globally. Yet the link to 'terrorism' of current 'bioterrorism', while certainly intense at the moment, is less novel than is often claimed.[38] We learn from chapters in this book that these concerns are often only apparently new, and are better conceptualized as recent manifestations of pre-existing clusters of modern concerns to do with territoriality, security and communicable disease. If European intervention into Ottoman territory to regulate the Mecca pilgrimage was 'pre-emption' as Zylberman argues, it was also 'bio-preparedness'. And we see from Stern's work that programs for health security have, often enough, squarely involved the military. Current policy emphasis on 'homeland' (that is territorial) security, needs to be understood as the latest expression of an enduring link between disease and national defense.

Conclusion

The chapters in this book detail various historically specific but returning logics and measures through which security from disease has been sought: programs of movement restriction and quarantine at local, national and global levels; programs of eradication by disinfection and vaccination; 'pre-emptive' public health action, including the implementation of primary health; the conflation of migration barriers and health barriers; surveillance of people and of information flow. Jointly, the chapters both draw and qualify a historical shift from absolute measures (of quarantine) to relative measures (of surveillance), from territory-oriented policing to network-oriented technologies, from old cordon sanitaires on sites of commercial exchange to a world where trade barriers double as lines of hygiene.[39] Inescapable is a sense of the intertwining of the national, the colonial, the international and the global in public health of the modern period. The national *was* the colonial, the colonial was already, the global. These spaces of regulation and movement did not emerge in a neat sequence, but as overlapping modern social and political formations. The interconnections between commerce, colonialism (military and cultural), national self-interest and a deeply politicized public health reveal the long past of current organizational, funding and discursive links between aid, development, foreign policy and disease management. This is both a historical and a historiographical point, suggesting the need further to link previously disparate literatures.

'Nothing can bring back the hygienic shields of colonial boundaries. The age of globalization is the age of universal contagion', write Hardt and Negri in their now famous book *Empire*.[40] Yet these collected chapters suggest a rather more complicated historical connection between colonialism, hygiene barriers, globalization and (what is missing in their formula) nationalism. In fact, we can profitably rearrange these elements. The age of colonization is better understood as the (first) 'age of universal contagion'.[41] We can see that nineteenth-century colonial boundaries were minimal compared to the exclusionary and segregative capacity of twentieth-century national boundaries. And in fact it was *lack of* regulation, the *absence* of 'hygienic shields' which most characterized the colonial world flow of people, goods, and disease. It was rather more the 'age of nationalism' which embedded 'hygienic shields' into border regulation. Further, the 'age of globalization' – the world post-HIV/AIDS, post-multidrug resistant TB, West Nile Virus and SARS – might be 'global' in terms of disease spread, but is also characterized by increasingly intense regulation at national borders. Medicine at the national border, indeed, is not really being 'brought back', it is spreading and deepening from places where it never went away. As SARS revealed, and as pandemic influenza may, medico-legal border control both haunts and challenges the trend towards transnational globalization.

Notes

1 M. Hardt and A. Negri, *Empire* (Cambridge Mass: Harvard University Press, 2001), p. 136.

2 H. Bell, *Frontiers of Medicine in the Anglo-Egyptian Sudan, 1899–1940* (Oxford: Clarendon Press, 1999), p. 4.

3 See F. Cooper and R. Packard (eds) *International Development and the Social Sciences: essays on the history and politics of knowledge* (Berkeley: University of California Press, 1997).

4 N. Goodman, *International Health Organizations and their Work* (New York: Churchill Livingstone, 1971); N. Howard-Jones, *International Public Health between the two world wars: the organizational problems* (Geneva: World Health

Organization, 1978); W.F. Bynam, 'Policing Hearts of Darkness: Aspects of the International Sanitary Conferences', *History and Philosophy of the Life Sciences*, 15 (1993): 421–34; J. Siddiqi, *World Health and World Politics: the World Health Organization and the UN System* (Columbus: University of South Carolina Press, 1995); K. Lee, *Historical Dictionary of the World Health Organization* (London: The Scarecrow Press, 1998); A.M. Stern and H. Markel, 'International Efforts to Control Infectious Diseases, 1851 to the Present' *Journal of the American Medical Association*, 292 (2004): 1474–9.

5 D.P. Fidler, *International Law and Infectious Diseases* (Oxford: Clarendon Press, 1999).

6 Stern and Markel, 'International Efforts to Control Infectious Diseases', 1476.

7 P. Weindling, *Epidemics and Genocide in Eastern Europe* (Oxford: Oxford University Press, 2000); I. Löwy and P. Zylberman, 'Medicine as a Social Instrument: Rockefeller Foundation, 1913–45', *Studies in the History and Philosophy of the Biological and Biomedical Sciences*, 31 (2000): 365–79.

8 M.D. Dubin, 'The League of Nations Health Organization' in P. Weindling (ed.) *International Health Organizations and Movements, 1918–1939* (Cambridge: Cambridge University Press, 1995) pp. 56–80; L. Wilkinson, 'Burgeoning Visions of Global Public Health: The Rockefeller Foundation, The London School of Hygiene and Tropical Medicine, and the "Hookworm Connection"', *Studies in the History and Philosophy of Biological and Biomedical Sciences*, 31 (2000): 397–407; L. Manderson, 'Wireless Wars in the Eastern Arena' in P. Weindling (ed.) *International Health Organizations and Movements* (Cambridge: Cambridge University Press, 1995).

9 J. Farley, *To Cast Out Disease: A History of the International Health Division of the Rockefeller Foundation* (New York: Oxford University Press, 2003); J. Gillespie, 'The Rockefeller Foundation and Colonial Medicine in the Pacific' in L. Bryder and D. Dow (ed.) *New Countries, Old Medicine* (Auckland: Auckland University Press, 1995); M. Cueto (ed.) *The Missionaries of Health* (Bloomington: Indiana University Press, 1994); C.J. Shepherd, 'Imperial Science: The Rockefeller Foundation and Agricultural Science in Peru, 1940–1960', *Science as Culture*, 14 (2005): 113–37.

10 See for example, M. Worboys, 'Manson, Ross and colonial medical policy: tropical medicine in London and Liverpool, 1899–1914' in R. Macleod and M. Lewis (eds) *Disease, Medicine and Empire: Perspectives on Western Medicine and the Experience of European Expansion* (London: Routledge, 1988); A. Marcovich, 'French colonial medicine and colonial rule: Algeria and Indochina' in Macleod and Lewis (eds) *Disease, Medicine and Empire*; A.M. Moulin, 'Tropical without the Tropics: The Turning-Point of Pastorian Medicine in North Africa' in D. Arnold (ed.), *Warm Climates and Western Medicine* (Amsterdam: Rodopi, 1996), pp. 160–80; A. Bashford, '"Is White Australia Possible?" race, colonialism and tropical medicine in the early twentieth century', *Ethnic and Racial Studies*, 23 (2000): 112–35; D.M. Haynes, *Imperial Medicine: Patrick Manson and the Conquest of Tropical Disease* (Philadelphia: University of Pennsylvania Press, 2001).

11 See M. Harrison, *Climates and Constitutions* (Oxford: Oxford University Press, 1999); W. Anderson, *Cultivation of Whiteness: Science, Health and Racial Destiny in Australia* (Melbourne: Melbourne University Press, 2002).

12 But see D. Armus (ed.) *Disease and the History of Modern Latin America: from Malaria to AIDS* (Durham and London: Duke University Press, 2003); W. Anderson, '"Where every prospect pleases and only man is vile": laboratory medicine as colonial discourse' in V. Rafael (ed.) *Discrepant Histories: Translocal Essays on Filipino Cultures* (Philadelphia: Temple University Press, 1995).

13 For the former, see R. Acheson and P. Poole, 'The London School of Hygiene and Tropical Medicine: A child of many parents', *Medical History*, 35 (1991): 385–408; D. Fisher, 'Rockefeller Philanthropy and the British Empire: The Creation of the London School of Hygiene and Tropical Medicine', *History of Education*, 7 (1978): 129–43.
14 Manderson, 'Wireless Wars in the Eastern Arena'.
15 But see see Stern and Markel, 'International Efforts'; R. Hankins, 'The World Health Organization and Immunology Research and Training, 1961–1974', *Medical History*, 45 (2001): 243–66.
16 D. Chakrabarty, *Provincialising Europe: postcolonial thought and historical difference* (Princeton: Princeton University Press, 2000).
17 See S. Amrith, *A New Utopia: International Health and the End of Empire in Asia* (London: Palgrave, 2006).
18 U. Baxi, 'Global Neighbourhood and Universal Otherhood' cited in O. Aginam, 'The Nineteenth-century Colonial Fingerprints on Public Health Diplomacy: A Postcolonial View', *Law, Social Justice and Global Development Journal*, 1 (2003); See also N.B. King 'Security, Disease, Commerce: Ideologies of Post-Colonial Global Health' *Social Studies of Sciences*, 32 (2002): 763–89; N.B. King, 'The Scale Politics of Emerging Diseases', *Osiris*, 19 (2004): 62–76.
19 N. Shah, *Contagious Divides: Epidemics and Race in San Francisco's Chinatown* (Berkeley: University of California Press, 2001); A.L. Fairchild, *Science at the Borders: Immigrant Medical Inspection and the Shaping of the Modern Industrial Labor Force* (Baltimore: Johns Hopkins University Press, 2003); A. Bashford, *Imperial Hygiene: a critical history of colonialism, nationalism and public health* (London: Palgrave, 2004), ch. 6.
20 See P. Zylberman in this volume; M. Harrison, 'Cholera theory and sanitary policy' in his *Public Health in British India: Anglo-Indian Preventive Medicine, 1859–1914* (Cambridge: Cambridge University Press, 1994).
21 H. Markel, '"Knocking out the Cholera": Cholera, Class and Quarantines in New York City, 1892', *Bulletin of the History of Medicine*, 69 (1995): 420–57.
22 Andreas Schloenharrdt details these Acts, but does not make this point in, 'Exclusion of Infected Persons under Immigration Laws in Asia'. Paper presented to the Infectious Diseases and Human Flows in Asia workshop, University of Hong Kong, June 2005.
23 Fairchild, *Science at the Borders*, passim.
24 Shah, *Contagious Divides*, p. 204.
25 Bashford, *Imperial Hygiene*.
26 Shah, *Contagious Divides*.
27 For the latter, see for example, H. Deacon, 'The Politics of Medical Topography: seeking healthiness at the Cape during the nineteenth century' in R. Wrigley and G. Revill (eds) *Pathologies of Travel* (Amsterdam: Rodopi, 2000) pp. 279–98; L. Bryder, '"A Health Resort for Consumptives": Tuberculosis and Immigration to New Zealand, 1880–1914', *Medical History*, 40 (1996): 453–71.
28 R. Coker, 'Compulsory screening of immigrants for tuberculosis and HIV: is not based on adequate evidence, and has practical and ethical problems', *British Medical Journal*, 328 (2004): 298–300; R. Coker, *Migration, public health and compulsory screening for TB and HIV* (London: Institute for Public Policy Research, 2003).
29 M. Griffiths and T. O'Callaghan, *International Relations: the key concepts* (London: Routledge, 2002), pp. 126–7.

30 J.A. Scholte, *Globalization* (London: Palgrave, 2000), p. 16.
31 For example, K. Lee, K. Buse, S. Fustukian (eds) *Health Policy in a Globalising World* (Cambridge: Cambridge University Press, 2002); G. Berlinguer, 'Globalization and Global Health', *International Journal of Health Services*, 29 (1999): 579–95; M.E. Wilson, 'Travel and the Emergence of Infectious Diseases', *Emerging Infectious Diseases*, 1 (1995).
32 A.G. Hopkins, 'The History of Globalization – and the Globalization of History?' in Hopkins (ed.) *Globalization in World History*, Pimlico, 2002, pp. 11–46; M. Geyer and C. Bright, 'World History in a Global Age', *American Historical Review*, 100 (1994): 1034–60; B. Mazlich, 'Comparing Global History to World History', *Journal of Interdisciplinary History*, 28 (1998): 385–95.
33 Le Roy Ladurie, 'The Unification of the Globe by Disease' in *The Mind and Method of the Historian* (Chicago: University of Chicago Press, 1981); I. Catanach, 'The "Globalization" of Disease? India and the Plague', *Journal of World History*, 12 (2001): 131–53; S. Watts, *Disease and Medicine in World History* (London and New York: Routledge, 2003); D. Igler, 'Diseased Goods: Global Exchanges in the Eastern Pacific Basin, 1770–1850' *American Historical Review*, 109 (2004); D. Arnold, 'The Indian Ocean as a Disease Zone, 1500–1950', *South Asia*, 14 (1991): 1–21.
34 But see R. Packard, P. Brown, H. Frumkin, R.K. Berkelman (eds) *Emerging Illnesses: Negotiating the public health agenda* (Baltimore: Johns Hopkins University Press, 2004); K. Loughlin and V. Berridge, *Global Health Governance: Historical Dimensions of Global Governance* (London: London School of Hygiene & Tropical Medicine and World Health Organization, 2002).
35 See P. Finney (ed.), *International History* (London: Palgrave, 2005). The absence of any discussion of international health organizations in this collection characterizes the current lack of integration of these fields.
36 This idea is developed in A. Bashford, 'Global biopolitics and the history of world health', *History of the Human Sciences*, 19 (2006): 67–88.
37 C. Loh and Civic Exchange (eds) *At the Epicentre: Hong Kong and the SARS Outbreak* (Hong Kong: Hong Kong University Press, 2004).
38 See for example, E. Fee and T.M. Brown, 'Pre-emptive Biopreparedness: Can We Learn Anything from History?', *American Journal of Public Health*, 91 (2001): 721–5; N. Yand and X. Wang, 'Disease Prevention, Social Mobilization and Spatial Politics: the Anti-Germ-Warfare Incident of 1952 and the "Patriotic Health Campaign"', *Chinese Historical Review*, 11 (2004): 155–82; R. Rogaski, 'Nature, Annihilation and Modernity: China's Korean War German-Warfare Experience Reconsidered', *Journal of Asian Studies*, 61 (2002): 381–415; B. Balmer, *Britain and Biological Warfare: Expert Advice and Science Policy, 1930–1965* (London: Macmillan, 2001); S.H. Harris, *Factories of Death: Japanese Biological Warfare, 1932–45 and the American Cover-up* (New York: Routledge, 2002).
39 Thanks to Claire Hooker for discussion on these ideas.
40 Hardt and Negri, *Empire*, p. 136.
41 For example, A. Crosby, *Ecological Imperialism: the biological expansion of Europe, 900–1900* (Cambridge: Cambridge University Press, 2004); K.F. Kiple and S.V. Beck (eds) *Biological Consequences of the European expansion, 1450–1800* (Ashgate, 1997); G.W. Lovell, '"The Heavy Shadows and Black Night": Disease and Depopulation in Colonial Spanish America', *Annals of the Association of American Geographers*, 82 (1992): 426–43.

Part I

World Health: National and Colonial Histories

2

Civilizing the State: Borders, Weak States and International Health in Modern Europe

Patrick Zylberman

In the nineteenth century, major epidemics raised border-related issues among people and animals. Cholera? The border between East and West. Cattle plague? The border between Tsarist Russia and Europe. Malaria? The frontier of modernization in the Balkans. The modern state has defined itself by its borders, whose complexity is a function of the state having become more complicated and organized.[1] Public health has actively participated in this increasing complexity. Quarantines, international health regulations and delousing stations altered certain state's relations with other states, as well as with its own society. France in 1822, Texas in 1910, or even the Germany of victorious bacteriology, are examples of modern health institutions spreading inward from the borders rather than outward from the center.[2]

In the aftermath of World War I, staking out borders seemed to be the indispensable condition for the birth or rebirth of nations.[3] Based on a system of public services (such as sanitary cordons, disinfection stations, surveillance and observation) and on special regulations applying to individuals (such as crossborder workers, migrants, travelers, pilgrims or refugees), the rationale of borders determined more than ever a state's health policy. From quarantines to epidemiological surveillance, this era of 'health borders' spawned two developments: surveillance of individuals by medical police and, more importantly, the medical control of people from nearby or distant countries. Medicine and hygiene became a full part of the 'standard of civilization' that still prevails in international law.[4]

The Hague Peace Conferences established this standard of civilization in 1899 and 1907; and the League of Nations (henceforth LN) reaffirmed it in 1920 during preliminary sessions for setting up a Permanent Court of International Justice. It was, of course, mainly the European powers that first conceived such a notion. Mid-nineteenth-century 'evangelical imperialism' was a harbinger of the juridical concept. Indeed, the London Missionary Society wanted to change 'not just the faith of Africans but also their mode of dress, hygiene and housing'.[5] And, following its victory over Russia in 1905, Japan's entry into the club of 'civilized' states made the latter somewhat less Eurocentric. But what exactly was this norm of 'civilization'?

First of all, civilization referred in this context both to a state's relation to individuals and its relations with other states. When the term first appeared at the end of the eighteenth century, 'civilization' suggested what had been called up to that time 'police': security, tranquility, but also civility. Civility is the result of education; security and tranquility the result of 'regular government'.[6] These are the themes that François Guizot discussed in 1828 when he defined the 'fact' of civilization by reference to two symptoms: 'the development of social activity as well as of individual activity, the progress of society and the progress of humanity'.[7] Among liberal jurists, that diptych was then transformed into a triptych. The activity of the state is oriented in three directions: the individual, the community, and the human species. According to a 1911 formulation, unless the 'police' of a state corresponds to this three-fold criterion, it cannot be considered as civilized.[8]

We know from Fernand Braudel that civilization simultaneously implies a will to exert influence over other societies and an unwillingness to borrow from others. A permanent tension between these two poles is inherent to any civilization.[9] The implication is that 'civilizing' a state does not refer to the gradual lessening of its use of coercion,[10] but rather to the history of the limits it has set for itself, legally and geographically.[11] What defines a health border reflects the considerable influence that geopolitical and cultural-geographical concepts were to have upon epidemiological policy. From this standpoint, delineating a health border is not simply designating the contours of a communicable disease on a map, but rather the sketching out of a system of defense against epidemiological hazards or, to put it better, against the terror of epidemiological hazards. Characteristically, as Alison Bashford has shown, health borders combine the language of epidemiology and medicine (epidemic, contagion, immunity) with the vocabulary of national defense (protection, invasion, security). Quarantine, for instance, has exerted a strong influence on 'the imagining of Australia as an island-nation, in which "island" stood for "purity" but also therefore vulnerability to invasion by infectious disease'. As such, Australia was conceived as a 'geo-body'.[12] But more than that, quarantine made Australia a 'civilized' state: to be considered civilized, it had to set up the territorial and administrative structures necessary for protecting its citizens against risks that, though of a biological nature, arose out of policies regarding health, quarantine, religion, and so forth, practiced by other states.

A 'civilized' state, therefore, is best described as a 'zone' or as a bounded space, with health boundaries defining not only its secondary but sometimes its primary border. Here, the historian joins the jurist. For the latter, the border, 'a zone of contact and relationships of contiguity between states', is administratively defined as a 'zone of public service' which is organized according to internal law but whose function requires cooperation with services in other governments.[13] From a health standpoint, its principle resided in direct cooperation between the states' health information agencies, a principle that was first proposed by the United States during the International Sanitary Conference in Washington in 1881,[14] as well as in the adoption of a series of formalities (medical examination, inspection, expulsion) whose goal was to prevent the importation of dangerous germs. More than a line of demarcation, the border took on the role of strategic keystone and watchtower for health.

Quarantines represented a liberal practice, unafraid to simultaneously affirm the sovereignty of individual rights and to rest on 'the power of norms and controls'.[15] In the name of freedom of traffic and exchange, the twentieth-century society of states worked to lift barriers to free movement which were imposed during the nineteenth century (the interruption of communication, interminable confinement or the military enforcement of quarantines). The norm of 'civilization' meant to 'civilize' the state, to 'civilize' the police and the government.[16] Rather than instructing people how to live, the International Health Regulations were, above all, an attempt to establish norms for administrative practices. Health borders had to be effective but also had to pursue the common good while respecting the rights of individuals, since the former is no excuse for the state harming some people in order to save all the others. Under cover of the protection and defense of individuals against health risks, this same 'civilizing purpose' would nevertheless result in an increase in the regulatory powers of states.[17] The 'complete efficacy' of the sanitary border was in fact sought and this could only be accomplished with a system of surveillance and of 'continuous verification' of individuals, and through 'restrictions on private individuals'.[18] The rise of individual rights, the strengthening of norms and controls: this contradictory and ambivalent movement defined the 'civilization' of the state. And a nation that had not met this standard might be sovereign, but was nonetheless regarded by others as a 'second-rate', weak or defective, state.[19]

According to Francis Fukuyama, 'weak or failed states are the source of many of the world's most serious problems, from poverty to AIDS to drugs to terrorism'.[20] Following the end of the Cold War in areas ranging from the Balkans to Caucasia, from the Middle East to South East Asia, weak or foundering states have caused many a humanitarian catastrophe. And 11 September 2001 has demonstrated that such states represent a strategic menace often discussed in terms of a 'civilized-uncivilized' continuum. Not simply related to national defense or the energy supply, these risks extend to infectious diseases, a 'nontraditional threat' to state security. According to the Institute of Medicine's *America's vital interest in global health* (1997), poor health and global insecurity are linked to economic underdevelopment and political instability.[21] The inability of states to fight against major epidemics or organize a health system might cause major sanitary as well as political disasters with a chain reaction of repercussions on social and political systems.

How to deal with faulty states? This was a key problem in the history of both national and international health on the European continent during the past two centuries. I will briefly examine the policy of health borders in three contexts where states were considered weak or defective by European powers. Examining the health border between the Ottoman Empire and the rest of Europe in the late nineteenth century, we shall see that pre-emption was the first policy for managing health borders with a weak state.[22] We shall then see how the quarantine model contributed to defining and consolidating national borders in Yugoslavia during the 1920s, when nation-building seemed to provide a method for consolidating weak states. The third context will be the LN's initial attempts to internationalize the health border between Poland and Russia in 1921–1923. In conclusion, I shall point out some implications of these three examples for the current crisis of the Westphalian system of international public health.

Pre-emption: The Ottoman Empire, cholera and the Muslim pilgrimage

The 'Westphalian' system of public health – a phrase derived from the Treaty of Westphalia in 1648, which brought the Thirty Years War to an end – was based on the principle of non-intervention in the domestic affairs of other states. The first international health convention that was actually implemented, signed in Paris in 1903, placed public health within this international framework. International health regulations, the essence of this Westphalian system, did not require that governments improve sanitary conditions in their homeland. The intention was to protect international trade and keep germs and communicable diseases from circulating across borders. This horizontal system involved only states and traditional interstate relations: diplomacy, trade and warfare.[23]

Diplomacy, trade and warfare: isn't this cholera from a political perspective? Consider the Belgian Consul General's understanding that international hygiene is nothing more than the struggle of 'various European governments against the progressive invasion ... of cholera'.[24] Since the cholera epidemic of 1865, Europe had found itself 'at the mercy of the pilgrimage to Mecca' which until then had been a threat only to Egypt and Syria.[25] What characterized the epidemic of 1865 was first that the disease 'traveled from the Indies to Europe *by only one pathway*, that of Mecca and Egypt'. This epidemic 'inaugurated the maritime pathway' as Proust put it.[26] In addition, it was transmitted not by ships' passengers, but rather by sailors, pilgrims, Asian or African laborers, immigrants; in a word by the poor, the tired, the malnourished.[27] The entire geography of disease, not only for Western Asia, but also for the world, was thus transformed, and, significantly, in a relatively short time.[28] This rapid dissemination, now that most of the pilgrims return by sea, was noted by all the authors, following the deadly epidemic wave.

The 1865 epidemic became a model; the first pandemic of the industrial age, the first *high tech* epidemic in history, and as such, also the first that turned out to be the paradoxical product of civilization's progress, specifically the increase in number and rapidity of travel by sea (steamships) and on land (railroads). The opening of the Suez Canal in 1869 intensified these factors. Bombay was closer to Trieste (traveling time was reduced by 63 per cent), to Marseille (59 per cent), to London (44 per cent), to Hamburg (43 per cent). Born of the terror sown by the epidemic of 1865, health questions thus became solidly anchored in security questions for the first time. The Europe of 1865 shared the fear of cholera.

Between 1868 and 1892, the number of pilgrims taking the sea route to Mecca from the Far East, the east coast of Africa, Egypt, and North Africa more than doubled.[29] At the same time, caravans, which in 1869 transported around 77 per cent of the faithful to the Holy Places, carried only 33 per cent in 1893.[30] At that time, Persia was the last place where land routes with the Hijaz persisted. As carriers of a double contagion, political on the one hand (brotherhoods, pan-Islamism) and health on the other (plague, cholera), the pilgrims aroused considerable anxiety.[31] Between 1866 (at which time restrictive measures were begun at the initiative of France) and 1907 (the ratification of the *Convention de Paris*), health security was rarely divorced from the desire of the major powers to obtain a 'slowing of religious movements' in Arabia.[32]

Without doubt, there was a health issue in the Eastern question. Or rather, there was pressure from geo-epidemiological realities on the perceptions of different actors concerning the stakes in the strategic theater.[33] It is little wonder then that administrative red tape, brutal detentions, obstacles of all kinds were placed in the path of pilgrims and travelers. To bureaucratic harassment were added the even more difficult and very harsh conditions of detention in the Red Sea region. A single and perfectly representative example will suffice: the lazaret at al-Tûr in the Sinai, administered by the Egyptians. Opened in 1866, the station (9 kilometers wide and 500 meters deep) operated concurrently with the lazaret at al-Wadjh (under Ottoman administration). Upon their return from Mecca, the pilgrims were landed there before their entry into Egypt, for a period of 15 to 20 days if a suspected case of cholera was declared on board. Consigned to that 'normally deserted beach', sometimes for 3 to 4 months (the duration of the sentence was extended following each new case), they suffered from excessive heat during the day and from nights of piercing cold.[34] The north wind blew sand, causing conjunctivitis and other suppurating eye conditions.[35] In 1887, it was noted that 'everything is lacking' in this waiting zone, it was 'barely possible to find what is strictly necessary for the destitute and the beggars'.[36]

The camp was composed of four groups of 60 tents, which were the collective property of the shipping companies and could shelter between five and ten persons. Traders were housed in a further half a dozen tents. Each group had two cisterns, sometimes able to be used only five hours a day, and since many of the pilgrims' containers were broken or burned by the administrators, many people could not keep a supply of water on hand.[37] 'They therefore go through the night tormented by thirst, amidst the cries of children also dying from thirst. This is indeed a situation which calls for compassion', one among them complained.[38] Rich pilgrims were able to substitute themselves at disinfection sessions by the poor, and health certificates were sold in Suez in exchange for bribes without any medical visit. In addition, it was not rare that the head camp physician took bribes from the ship captains to shorten the duration of the detention.

Quarantine was also a militarized space. In 1885, 200 men guarded the encampment; 5 years later, the pilgrims were guarded by 450 soldiers.[39] Similarly, in order to prevent escapes, the Health Council in Alexandria decided that the pilgrims' ships would be escorted from Suez to Port-Said by a double row of armed guards. These included soldiers in dinghies, platoons on camel back along the banks, and militarized vigil-boats stationed around the ships during their stay at anchor in Suez.[40] The order was given to shoot anyone trying to escape. In a word, a ship owner in the Red Sea wrote that the quarantines were nothing more than a 'tyranny tempered by *bakchik*'.[41]

Such rigorous quarantine measures in the eastern part of the Mediterranean stayed in effect well into the twentieth century. The International Sanitary Convention ratified by 34 states in July 1933 would mark the turning point.[42] Why did this take so long? Although European liberals increasingly opposed the principle of quarantine within Europe, they endorsed quarantine measures elsewhere, which, in their opinion, had the purpose of 'isolating Europe from all dealings with the uncultivated and unhealthy habits of Oriental populations'.[43]

In 1866, the Sanitary Conference in Constantinople set 10 days as the period of quarantine for the three main pestilential diseases: plague, cholera and yellow fever. In 1903, the Sanitary Conference in Paris made a major change. On the basis of improved knowledge about the transmission of these pathologies rather than on the supposition of a uniform period of incubation, specific measures were introduced for each disease.[44] The general acceptance of the germ theory by medical circles at the turn of the twentieth century brought an end to the drawn-out debate on quarantines, a method that would be replaced with disinfection and measures for the surveillance of ships and passengers.[45] As we can see, progress in defining regulations – in this instance, the repeal of measures for detaining patients or those in contact with them – was closely linked to the discovery of the complex transmission of disease.

If strict quarantine was gradually reduced, other coercive measures were innovated under new surveillance systems. Stool examinations counted among the highly controversial measures imposed at the time. In 1911, at al-Tûr, they were mandatory for all pilgrims, whether or not they had symptoms of cholera. In May 1926, the International Office of Public Hygiene questioned the need for these procedures, invoking both epidemiological reasons and the 'intolerable restrictions' on the free circulation of people. Finally, in 1951, article 69 of the International Sanitary Regulations would state: 'No one can be forced to submit a rectal sample'.[46] But there were other coercive rules. Starting in 1926, International Health Conventions recommended that all pilgrims be vaccinated against cholera, a regulation which had been adopted by Egypt in 1902,[47] even though the vaccine's efficacy was still being debated. In 1956, given improvements in sanitary conditions during the pilgrimage, the special conditions regarding pilgrims in the Sanitary Regulations of 1951 were repealed;[48] and the WHO's Ninth Assembly adopted an additional regulation for pilgrims, who were now likened to migrants and seasonal workers as 'persons taking part in major meetings'. This moderate loosening of regulations was a response to criticism that was not only a matter of law but also of bacteriology and preventive medicine.

Ideas about introducing a more liberal policy on quarantine had been floated since the 1880s. At the 1887 International Hygiene Congress in Vienna, Émile Vallin had suggestively declared that international health regulations should be understood in the same spirit as the 1864 Geneva Convention.[49] In the same way that the Red Cross did not try to determine whether a war was just or unjust, health regulations should not pronounce whether states have the right to set up embargos or quarantines. At the time, such regulations were intended to lay down norms and rules of conduct for the services that organized and supervised detentions. They had to do with what Fukuyama has called the state's strength (in contrast with its scope), namely: 'the ability to formulate and carry out policies', 'to administrate efficiently', 'to control graft, corruption, and bribery', 'to maintain a high level of transparency and accountability', and 'to enforce laws'.[50] In a word, this was the 'civilization' of local administration.

The gradual decline of quarantines brought attention to bear precisely on the local conditions that bred contagion: in the case of the pilgrimage, the Sultan's health administration. While Constantinople's pan-Islamic ideology enabled it to exercise influence over political and religious aspects of the pilgrimage,[51] it was much too weak to block out cholera. The Sultan's health administration was shortsighted and unable to raise the population's concern for hygiene.[52] Little wonder that, when Ernest Renan made the case for not including Islam in the civilized world in 1862, he was applauded by physicians.[53] Even though Constantinople might master public health techniques (medicine and hygiene being the true missionaries from the West), core values such as solidarity, individualism and scientific freedom remained alien to Ottoman culture. Then as now, the question loomed of how to spread efficient state institutions despite economic, technological, political and cultural differences.

Interest in the European sanitary order was not only imposed on, but occasionally invited by, the Ottoman Empire. Twice, in 1838, when it established the Conseil Supérieur de Santé in Constantinople,[54] and in 1866, when it organized the international conference on cholera, the Ottoman Empire asked the European powers to let it be a part of their sanitary domain. However, the Ottoman Empire resisted Western European health proposals more than it complied with them. In 1893, it set as a condition for ratifying the Vienna Sanitary Convention that Ottoman sovereignty over health institutions and regulations should be maintained throughout its domains.[55] This demand, which had already been raised during the 1866 conference, led to a diplomatic standoff. Successive governments in Constantinople did not appreciate an international administration controlling the organization of public health in areas of Europe under the Sublime Porte's sway. In 1912, the Empire finally obtained recognition of its sovereignty over the Muslim pilgrimage, which it took to be a state issue and not a mere 'police' matter.[56]

But did the Ottoman Empire actually have the institutional capacity to implement health security measures? Following the Conference of Constantinople, Istanbul took the 'difficult' decision to send a medical commission to the Hijaz each year during the hajj.[57] A Health Office directed by an Egyptian physician was established at Jidda to watch over the port, the ships and the pilgrims, and to give free care to the sick. Yet, the greatest skepticism reigned among the European consuls on the ability of the Ottomans to enforce quarantine regulations: in Jidda, the Office was in reality limited to a single physician whose mission, nonetheless, was to centrally administer all the health posts.[58] Hospital capacity was also lacking, as much in Mecca (where the few mediocre hospitals dated from Muhammad Ali) as in Jidda where only one military hospital had existed since 1869. Moreover, the Ottoman authorities had never communicated the slightest reliable statistics on the pilgrimage; in fact, they were never collected.[59] And how could they be? Without vital statistics or surveillance of burials there was freedom to indulge in falsification and fraud. Neither arrivals nor departures were recorded in the hospitals; patients and deaths were counted together. What followed was merely pretense. For all these reasons, it was not prudent for European powers to let Constantinople head key institutions for monitoring travel in the Mediterranean.

Most European authorities agreed that the defense of health in Europe was played out in the Mediterranean region. This was the French geo-epidemiological axiom, from Charles Nicolle to Maximilien Sorre. Seen from Paris, the struggle against cholera boiled down to 'identifying the places which must be strengthened against the invasion of this epidemic, and which may be considered as veritable *strategic positions*'.[60] Thus, to the north, prevent the contamination of the Black Sea and to the south, organize the Red Sea.[61] Holding forth in 1943 on 'the protection of health in the West', André Siegfried found nothing better than to wisely copy Fauvel and Proust. 'The Mediterranean, door to the Orient, door to the desert, door to the oceans, is a sort of step, more exposed to contamination than is Europe. It is the first threatened, the first affected, the first therefore which should be protected'.[62] Out of 14 sanitary conferences held between 1851 and 1938, nine included on their agendas questions of public health raised by the pilgrimage. A tenth meeting was specifically devoted to the topic in January 1929 (Beirut) and October 1930 (Paris). In the age of cholera and for the united Powers, internationalizing public health meant not only taking away the 'window on the West'[63] (Egypt) from the Porte, but also the surveillance and control of Islam at its center (the Hijaz and the Holy Places), between African and Indian Islam. The Conference of 1911 (Paris) provoked an Ottoman delegate to remark bitterly that 'the Conference had decided that no distinction would be made between countries in Europe and outside of Europe. But in the new version of the convention, there is a title relative to the "Special measures in Eastern and Far Eastern countries"'.[64] As the French Foreign Ministry already noted in 1909, Constantinople was, in fact, under constant European oversight in health matters.[65]

The actual place of surveillance and control, therefore, was by no means a minor issue. The European health system made a distinction between sovereignty and sphere of influence. From the beginning, the principles did not vary; control of navigation, surveillance of the pilgrimages, protection of Egypt.[66] The idea had taken shape that an expanding 'civilization' has a right of pre-emptive intervention when it runs risks posed by the powerlessness and dereliction of patrimonial states with insecure borders.[67] With cholera then, Islam had become the stumbling block to a new European sanitary order. At the same time, with the right to pre-emptive intervention and the isolation of Europe, the health imperative gave a principle of legitimacy to the demarcation of the continent. Public health thus participated in the definition of Europe's frontiers.

Nation-building: Malaria in Macedonia and the Kingdom of the Serbs, Croats and Slovenes

Pre-emptive intervention was not the only way to cope with weak states. An alternative was nation-building. According to Fukuyama, nations cannot be deliberately constructed; they arise out of history, not of design.[68] Nonetheless, deliberate policies have been used to shape national identities. Nascent Yugoslavia provides a case in point.[69] For Andrija Stampar, from 1919 to 1931 the Director of Health Services in the newly born Kingdom of the Serbs, Croats and Slovenes, public health was the 'most important chapter in national political life'.[70] Two major scourges were afflicting his country: tuberculosis and malaria, the latter a disease related directly to the question of borders.

Malaria was rife in areas of anarchic land use and uncontrolled deforestation, where populations with variable immunity tended to settle. This was the case in Macedonia, most of which Serbia had annexed from Bulgaria. Repeating the old Ottoman policy of colonizing border zones, Belgrade encouraged nearly 50,000 colonists to settle in Macedonia from 1920 to 1936. These settlers were native Macedonians, Croats, Slavs from Greece, Hungary or Italy, and, above all, Serbs – Balkan War veterans, border guards or *komitadjis* (members of the armed gangs who had sown terror in Macedonia). Several hundred thousand hectares confiscated from their former Ottoman owners, or taken from lands depopulated by 10 years of warfare, were given to these settlers.

Anti-malarial efforts in the province were concentrated along the borders with Greece, Albania and Bulgaria. Health centers, some rudimentary, others larger and better equipped, staked out a line of demarcation. In 1929, authorities, aided by the Rockefeller Foundation, carried out the first large-scale experiment in the region with 'greening'. An entire valley where 25,000 people lived in 36 villages and a small town along the Bulgarian border and where malaria was endemic were treated with Paris green, an arsenic-copper pesticide against mosquitoes. On the other side of the border, Bulgaria carried out similar projects. The Yugoslav Health Service had been conducting sanitation work in border villages since 1924. These local prevention and sanitation campaigns were to contribute directly to raising the peasantry's awareness of hygiene.[71]

This micro-level acculturation of the population to hygiene was also intended to help form a national conscience. According to C.E.A. Winslow, public health thus appeared for the first time as a central element in nation-building in the history of the Balkans.[72] But we must not confuse nation-building with nationalism.[73] Regardless of how sincere Belgrade's attempt to raise the level of public health in Macedonia might have been, it could never have succeeded without the technical and financial support of the League of Nations and the Rockefeller Foundation. From this perspective, nation-building was nothing less than a 'distinctive form of imperial tutelage'.[74] The Rockefeller Foundation's constant policy of handing over health institutions to host country authorities as soon as possible showed that the primary objective was to boost the government and public administration rather than to provide services to end-users – contrary to the present-day practices of non-governmental organizations [NGOs].

Other forces were also at work in the fight against malaria in Macedonia. The health system there, under the nationalistic impetus of Serbian settlers, was organized in a tight pattern of health stations along the province's borders. This 'health border' separated Yugoslav from Greek and Bulgarian Macedonia; it was a substitute for an indefinable ethnic borderline. This sanitary cordon laid down a border that wars and migrations had blurred. But the trick was to make these territorial boundaries 'natural' by using the quarantine line as a defense against endemics and epidemics, as markers of a 'biological self'[75] – of the nation as a biological entity. Not unlike the 'biologically justified border' to which German *Geopolitiker* were aspiring,[76] the health border tended to become a 'natural border' and, like the latter, justify recourse to force. The intention was not just to close gates at the border but also to set up a health network for protecting so-called ethnic customs and traits, a sort of racial quarantine similar to the quotas under the 1924 American Immigration Act.[77] Anti-malarial policy in Macedonia thus became a key factor in the invention of national identities in border areas. For Yugoslav hygienists, sanitation campaigns in villages were 'identified with high-level state policies'.[78] This explains why Belgrade put so much effort into equipping Macedonia with health stations and why the Yugoslav health system gravitated toward the country's periphery.

Internationalization: The 'sanitary zone' in eastern Poland (1921–1923)

In the case of the Ottoman Empire, the demand for health institutions came from outside; in the case of Yugoslavia, from inside. Another pressure was the epidemiological situation related to the redrawing of the map of eastern Europe after World War I. Peace came in 1919 in the midst of an epidemic in eastern Europe: 30 million cases of typhus, dysentery, malaria, tuberculosis and cholera; 3 to 3.6 million deaths between 1914 and 1923 in European Russia alone.[79] From 1918 to 1922, over the entire territory but mainly within the European part of the country, typhus (37 million cases), relapsing fever (27 million cases), malaria (23 million cases) and cholera (more than 1 million cases) killed unceasingly. Contamination passed from soldiers to refugees displaced by the war and the civil war. Epidemics over a period of 4 years significantly and horrifyingly added to the famine, which lasted only 1 year (1921–1922).[80] The figures from the period show a similar progression in other countries of eastern Europe.[81] In January 1920, 100,000 cases of typhus (12,000 deaths) were recorded in Poland; in December, Czechoslovakia and Hungary were struck in turn.[82] In early 1921, the incidence was subsiding. In Poland, the number of cases of typhus declined from 157,000 to 48,000, in Romania from 46,000 to

4,800.[83] A new deterioration of the situation took place during the winter of 1921–1922 under the effects of the famine which was raging in Russia, causing around 4 million cases in Poland.[84] Moreover, to typhus was added dysentery (15,000 cases in Poland in 1919, but 30,000 in 1920, especially in the regions invaded by the Red Army), typhoid fever, as well as a severe resurgence of smallpox during the last phase of the Russo-Polish war.[85] In Bohemia, too, morbidity rocketed, with the number of cases of smallpox climbing from two in 1913 to 1,181 in 1918 to 6,412 12 months later.[86] The Medical Director of the LN's Epidemic Commission was convinced that there would be no return to civilization and social progress without health security in eastern Europe.[87]

Despite the wave of epidemics, nation-states and people did not cooperate. The common danger did not arouse a joint-defensive response, nor did the general feeling of mutual vulnerability spur a reaction of solidarity. According to Wickliffe Rose, Director of the Rockefeller Foundation's International Health Board, 'the spread of contagious and deficiency diseases is but an element in an all-pervading complex which one may call the ills of Europe.' The War – not the Peace, as Keynes would have it – had thoroughly fragmented, both economically and politically, the east and the center of the Continent. The field of intervention in infectious diseases was so broad that, in order to deliver the new Europe from its torment, only a large-scale counteroffensive could succeed – a Koch-style campaign that, zeroing in on parasites in all human carriers, would be freed from the constraints of national boundaries and centrally planned by an international technical agency.[88]

Assembled in Warsaw from 20 to 28 March 1922, the first European Sanitary Conference held after the War set the goal of designing the tools for this global counteroffensive. Criticizing the shortcomings of sanitary cordons, it proposed instead a 'sanitary zone', an epidemiological and legal concept. As an epidemiological concept, it referred to a surveillance zone formed by a permanent system of hygiene institutes, bacteriological laboratories, quarantine stations and quarantine hospitals, that was positioned along a band 150 kilometers wide on each side of the Polish-Russian border.[89] As a legal concept, this zone corresponded to an area with mixed jurisdiction under international oversight, a zone that, though astride the political border between Poland and Russia, did not deny that border's legal validity.[90]

Did this signal a turning point in the history of health borders? Throughout the nineteenth century, the health frontier in Europe had evolved from an absolute (quarantine) to a relative (surveillance) border. The aim was no longer to stop infiltrations but rather to keep an eye on the flow. To be sure, the defense zone set up in eastern Poland was also part of a 'sanitary cordon', stretching from the Baltic to the Black Sea, against the Bolshevik contagion. But it also formed a medical no man's land, where states and people could exist next to each other without interacting.

Of Italian and German origin,[91] the idea of a health border zone was based on political geography's definition of three spaces.[92] A border was more than a simple line. It juxtaposed two opposite zones, each under the control of a different nation-state, and an in-between zone of fusion that came out of the 'development of the border within the border area'.[93] This region, which has been more appropriately called a 'zone of differentiation',[94] is a marginal interstice that is regularly a matter of dispute and is not under the permanent control of any bordering state. The 'sanitary zone' institutionalized this 'ambivalence'.[95] It tried to channel the flow of people not within the 'state-people-territory' framework used for national defense but, instead, by temporarily and locally suspending established national procedures for the sake of a more functional approach. It exemplifies an attempt to handle problems by 'debordering' them.[96] Barging into the territorial framework, the principle of functionality and international control transcended territory but without any attempt to lay down new lines of demarcation.

The codification of such a sanitary zone by international conferences in 1897, 1903 and 1922 was part of an effort to keep epidemics from spreading due to major population movements (cholera from Mecca, typhus from Smolensk or Minsk). In like manner, the USSR's collapse has re-aroused fear lest resistant strains of tuberculosis and HIV overwhelm Europe. The derelict Soviet bloc with its failing health institutions,[97] has become the image of 'passive regions' and 'drifting states', which are unable to meet the 'standards of the civilized state': internal surveillance of transportation infrastructures, border security, and a centralized administration.[98] This has justified the right of pre-emptive intervention claimed by Europe and the United States during the period of transition out of Communism.

Conclusion: The Decline of the Westphalian System of Public Health?

This claim to a right of pre-emptive intervention is deeply embedded in the past. Sociologically and culturally, an epidemic is both an outbreak of a new event and a return to old patterns of behavior.[99] Despite globalization and its economic impact in the context of real-time communications and of the supremacy of a worldwide organization of health over the principle of state sovereignty, was not Severe Acute Respiratory Syndrome (SARS) curbed in the summer of 2003 thanks to the 'medieval methods'[100] of quarantine and compulsory isolation of the sick? This pandemic made it difficult clearly to distinguish between the sanitary, economic and political interests of the states concerned.

The same was true in the era of cholera. For the Great Powers, as seen above, internationalizing public health meant taking Egypt away from the Sultan in Constantinople. It also implied a surveillance and control of the center of Islam, of the Hijaz and shrines. Was all this dictated by strategy or by prophylaxis? Adrien Proust, a health service inspector and France's representative to international health conferences, could barely tell the difference:

> It does not rest with us to tackle political questions; but, connected to permanent endemic sources by a continuous current of exchanges necessarily leading to Suez and to the narrow bottleneck of the canal opening into the Mediterranean [...] the immense Nile valley feeds into Africa, the center of Europe's colonial cupidity, movements of population [that have as a] first result to favor the expansion of Islamism and, consequently, access to the pilgrimage to Mecca.[101]

For the Great Powers, health surveillance of the Hijaz could not be separated from political control over the pilgrimage.[102] As John Gaddis has rightly pointed out, 'preemption sounds new only because it's old'.[103] On the one hand, international health was in the habit of clinging to the classical Westphalian scheme of things, in the sense that states remained the central actors of the system; on the other hand, since infringement upon the sovereignty of eastern lands was such a usual state of affairs within the Mediterranean region, it ignored the Westphalian concept in practice.

According to Fidler, the history of governance of infectious disease threats from the mid-nineteenth century to the beginning of the twenty-first century evolved as a series of institutional arrangements: the International Health Regulations based on non-interference and disease control at the border; the WHO's vertical programs based on disease control or eradication at the source and expert guidance; and finally the fall of the 'classical system' (IHR) leading to a new, post-Westphalian regime standing beyond state-centrism and national interest. Yet international public health cannot be reduced to Westphalian strategies.

Since 1865 competing poles of inspiration, institutions and actors have enriched global public health history. Three different strategies epitomized the changing patterns of international public health. Pre-emption was the dominant strategy in the eastern Mediterranean since the 1860s. It was based on the idea that adoption of the 'standard of civilization' (individualization of medical police and legal normalization of health controls) by eastern Mediterranean governments would upgrade what Fukuyama has recently called the state's strength, or the efficiency of the states in formulating and implementing their policies. While pre-emption had implications for the continuously expanding hegemony of the Great Powers, prevention on the contrary demanded an international cooperative framework. Spurring national identities, prevention helped international public health move towards a national health order rather than towards a supranational health order. Thus, preventing the spreading of disease through nation-building was a second strategy. A third was exemplified by the 'sanitary zone' positioned along the Polish-Russian border, in order to cope with the devastating epidemics that sprang up after World War I in eastern Europe. The 'sanitary zone' tried to channel the flow of refugees, not within the state-people-territory framework, but instead, by temporarily and locally suspending national laws for the sake of a more functional approach transcending the territory. Yet there was no attempt to lay down new lines of demarcation: it tried to handle problems by 'debordering' them.

In sum, the history of health borders highlights two themes: the one epidemiological (border diseases ranging from cholera through typhus and malaria to SARS) and the other political (the strategic remedies for coping with weak or derelict states, ranging from pre-emptive intervention through nation-building to the internationalization of sanitary borders). These two themes converge in the geo-epidemiological mentality of today's experts and leaders. The switch from a risk to a threat has taken us from a perception of international public health mainly as a technical matter,[104] to a conception of public health policy grounded, above all, in geopolitical ideas and in terms of national interests. Although one nation's health security depends on all others', national (or sectional) interests still prevail in public health.[105] This focus on microbial threats signals a return to the sources of a neo-quarantine policy, based on a revived 'norm of civilization' encompassing both public health and security.

Crucial issues are at stake here. How did the notions of sovereignty and security come to be seen as divisible in these complex epidemiological and political emergencies? Did the rising power of international and transnational organizations actually bring the legitimacy of states into question? The ending of the twentieth century did not seem to substantiate the post-Westphalian idea. The focus on microbial threats has led to rekindling borders during emergencies (the 'mad cow' crisis), and pre-emptive interventions in defective states have become a key element in international health politics (SARS). The rise and fall of Westphalian public health governance is only part of the full story.

translated from French by Noal Mellott (CNRS, Paris, France)
and Jon Cook (CERMES, Paris, France)

Notes

1 J. Brunhes & C. Vallaux, *La Géographie de l'histoire. Géographie de la paix et de la guerre sur terre et sur mer* (Paris: Alcan, 1921), pp. 338–9, 344, 363.

2 On the royal ordinance of 3 March 1822, see H. Rey, 'Les quarantaines', *Archives de médecine navale*, 22 (1874): 126–7; A.M. Stern, 'Buildings, boundaries, and blood: Medicalization and nation-building on the U.S.-Mexico border, 1910–1930', *Hispanic American Historical Review* 79 (1999): 54; P. Weindling, 'A virulent strain: German bacteriology as scientific racism,

1890–1920' in W. Ernst & B. Harris (eds), *Race, science and medicine, 1700–1960* (London: Routledge, 1999), pp. 218–34.

3 A. Stampar to L. Rajchman, 10/3/1924, League of Nations C.H./S.C./Malaria/2, League of Nations Archive, Geneva.

4 G.W. Gong, *The Standard of 'Civilization' in International Society* (Oxford: Clarendon Press, 1984).

5 N. Ferguson, *Empire: The rise and demise of the British world order and the lessons for global power* (New York: Basic Book, 2002), pp. 100–1.

6 E. Benveniste, 'Civilisation. Contribution à l'histoire du mot', *Problèmes de linguistique générale* (Paris: Gallimard, 1966), pp. 339, 343, 345.

7 F. Guizot, *Histoire de la civilisation en Europe* (Paris: Hachette, 1985), p. 64.

8 G. Jellinek, *L'État moderne et son droit: t. I – Théorie générale de l'État*, French transl. by G. Fardis (Paris: Giard & Brière, 1911), p. 410.

9 F. Braudel, *La Méditerranée et le monde méditerranéen* (Paris: Colin, 1966), vol. 2, pp. 101–9.

10 Norbert Elias's 'civilization process' confuses administrative powers with political authority and obstinately tries to assess changes in power (which is taken to be a pure social power) without taking into account politics. For, whether boundaries are health borders or not, we should bear in mind that 'there is no border save political borders'. P. Geouffre de Lapradelle, *La frontière* (Paris: Éditions Internationales, 1928), p. 11.

11 Jellinek, *L'État moderne*, p. 373; and Jellinek, *L'État moderne et son droit: t. II – Théorie juridique de l'État* (Paris: Giard & Brière, 1913), pp. 44–5.

12 A. Bashford, *Imperial Hygiene: A Critical History of Colonialism, Nationalism and Public Health* (London: Palgrave, 2004), p. 116.

13 Geouffre de Lapradelle, *La frontière*, pp. 15, 232, 262, 282–3.

14 A. Baarkus, 'The Sanitary Conferences', *Ciba Symposia* (1943), p. 1574; this was approved by twelve of the states attending the Conference, among whom Austria-Hungary, France, Italy, the Low Countries, Russia, and disapproved by five, including the United States, Great Britain and Turkey. *Protocoles*, Rome (1885), annex V to the Protocole no. 1, 63. It was not before 1907 (1911 in practice) that such a system would come into operation thanks to the creation of the Office international d'hygiène publique.

15 C. Lefort, 'les Droits de l'homme et l'État-providence', in *Essais sur le politique XIXe–XXe siècles* (Paris: le Seuil, 1986), p. 35.

16 P.E. Carroll, 'Medical Police and the History of Public Health', *Medical History*, 46 (2002): 461–94.

17 Jellinek, *L'État moderne*, p. 374.

18 Geouffre de Lapradelle, *La frontière*, pp. 15, 231, 262.

19 Jellinek, *L'État moderne*, p. 411.

20 F. Fukuyama, *State-building: Governance and world order in the twenty-first century* (London: Profile Books, 2004), p. ix. Thérèse Delpech, too, has emphasized the 'phenomenon of the state's failure' and the 'challenge that these chaotic formations represent for the whole of the organized world' in *Politique du chaos: L'autre face de la mondialisation* (Paris: Seuil, 2002), pp. 9–12.

21 N.B. King, 'Security, disease, commerce: Ideologies of post-colonial global health', *Social Studies of Science*, 32 (2002): 770, 781.

22 J.L. Gaddis, *Surprise, security, and the American experience* (Cambridge, MA: Harvard University Press, 2004), pp. 16–22.

23 D.P. Fidler, *SARS, Governance, and the Globalization of Disease* (London: Palgrave, 2004), pp. 23–35.

24 E. Sève, Sanitary Conference held in Washington (1881), in *Les dix premières années de l'Organisation mondiale de la Santé* (Genève: O.M.S., 1958), p. 14.
25 A. Proust, *Essai sur l'hygiène internationale. Ses applications contre la peste, la fièvre jaune et le choléra asiatique* (Paris: Masson, 1873), p. 45.
26 A. Proust, *la Défense de l'Europe contre le choléra* (Paris: Masson, 1892), p. 78.
27 J. Girette, *La civilisation et le choléra* (Paris: Hachette, 1867), pp. 144–6; Bartoletti, Rapport, *apud* A. d'Avril, *l'Arabie contemporaine* (Paris: Maillet, 1868), pp. 301–3.
28 Dr Catelan, 'Choléra au Hedjaz en 1890', *Travaux du Comité consultatif d'hygiène publique de France*, 21 (1891): 839.
29 Dr P. Alix, 'Contribution à la géographie médicale: Djeddah, pèlerinages, choléra', *Archives de Médecine navale*, 62 (1894): 327.
30 P. de Ségur Dupeyron, 'la Caravane de la Mecque', *Revue des Deux Mondes*, 10 (1855): 351, 358; Dr F. Duguet, *le Pèlerinage de la Mecque au point de vue religieux, social et sanitaire* (Paris: Rieder, 1932), pp. 135, 158.
31 This is a strongly engrained commonplace idea, as shown in the posthumous work by A. Siegfried, *Itinéraires de contagions. Epidémies et idéologies*, with a preface by Pasteur Valléry-Radot (Paris: Colin, 1960).
32 French Ministry of Foreign Affairs [henceforth MAE] CCC Djedda, vice-consul (Dubreuil), 3 March 1869.
33 P. Baldwin, *Contagion and the State in Europe 1830–1930* (Cambridge: Cambridge University Press, 1999), pp. 226–36.
34 Catelan, 'Choléra au Hedjaz en 1890', p. 835.
35 M. Delarue, *Rapport médical de M. Delarue, commisionné par le gouverneur général de l'Algérie à bord du 'Pictavia', pour le pèlerinage de la Mecque en 1891* (Alger: Giralt, 1892), p. 50.
36 Dr Abbate Pasha, *la Question de la neutralisation quarantenaire du canal de Suez devant le Congrès de Venise* (Le Caire: Imp. Baudieri, 1891), p. 17.
37 M.M.H. Farâhâni, *A Shi'ite Pilgrimage to Mecca 1885–1886*, translated and edited by H. Farmayan and E.L. Daniel (Austin: University of Texas Press, 1990), p. 290.
38 H. Lammens, le Pèlerinage de la Mecque en 1902. Journal d'un pèlerin égyptien, *Bull. mens. missions belges de la Compagnie de Jésus. Congo, Bengale, Ceylan*, 6, no. 2 (February 1904): 50; letters sent to *al-Ahram* by an Egyptian pilgrim.
39 Farâhâni, *A Shi'ite Pilgrimage*, pp. 289–90; Delarue, *Rapport médical*, p. 99.
40 Catelan, 'Choléra au Hedjaz en 1890', p. 837; Proust, *la Défense de l'Europe*, pp. 302–7.
41 Boissevain, manager of the ship company Nederland, quoted in Dutrieux Bey, *le Choléra dans la Basse-Egypte en 1883. Relation d'une exploration médicale dans le delta du Nil pendant l'épidémie cholérique* (Paris: Berthier, 1884), p. 251.
42 J. Gerlitt, 'The development of quarantine', *Ciba Symposia*, 2 (1940): 579. Health regulations of Mecca pilgrimages are analyzed in Duguet, *le Pèlerinage de la Mecque*, p. 122; and N.M. Goodman, *International health organizations and their work* (London: Churchill Livingstone, 1971), pp. 66–74.
43 Girette, *La civilisation et le choléra*, p. 35.
44 H.S. Gear & Z. Deutschman, *Transports internationaux et protection de la santé publique: La réglementation sanitaire internationale* (Geneva: World Health Organization, 1956), p. 13.
45 Anon., review of A. Chantemesse & F. Borel's *Frontières et prophylaxie (hygiène internationale)* (Paris: Doin, 1907), *Archives de médecine navale*, 87 (1907): 217–20.

46 R. Pollitzer, *Le choléra* (Geneva: World Health Organization, 1960), pp. 1026–7.
47 *Ibid.*, pp. 1021, 1027.
48 Gear & Deutschman, *Transports internationaux*, p. 57.
49 É. Vallin quoted by Anon., 'Congrès international d'hygiène de Vienne', *Archives de médecine navale*, 49 (1888): 61, 65. Émile Vallin (1833–1924) was chief editor of the *Revue d'hygiène*, which he had founded in 1879. He held the Chair of Hygiene and Legal Medicine in the army's École d'Application du Val-de-Grâce (1874).
50 Fukuyama, *State-building*, p. 12. See, too, N. Ferguson, *Colossus: The price of America's empire* (New York: Penguin Press, 2004), p. 197.
51 C. Snouck Hurgronje, 'Politique musulmane de la Hollande', *Revue du monde musulman*, 14, (1911): 450–2, 454–8.
52 Quoted by B. Lewis, *Europe-Islam, actions et réactions* (Paris: Gallimard, 1992), p. 19.
53 E. Renan, 'De la part des peuples sémitiques dans l'histoire de la civilisation' (1862) in E. Renan: *Qu'est-ce qu'une nation?* (Paris: Pocket, 1992), pp. 188–9, 198; C. Daremberg, *Journal des débats*, 16 August 1866, quoted in Girette, *La civilisation et le choléra*, p. 76.
54 On the Conseil Supérieur de Santé in Constantinople, see D. Pansac, 'Tanzimat et santé publique: les débuts du Conseil sanitaire de l'Empire ottoman' in D. Panzac, *Population et santé dans l'Empire ottoman, xviii*[e]*–xx*[e] *siècles* (Istanbul: Isis, 1996) pp. 77–85; and on the parallel Conseil quarantenaire of Alexandrie, see C. Essner, 'Cholera der Mekkapilger und internationale Sanititätspolitik in Aegypten, 1866–1938', *Welt des Islams*, 32 (1992): 41–82.
55 MAE series C/18/4, vol. 22, Barrère to Develle, 14 February 1893.
56 Dr C. Izzedine, 'Procès-verbaux de la Conférence sanitaire internationale de Paris (1912)', MAE, Fol. LF 128–99 (1), f. 80ff., 122, 444; also Clemow, the British delegate, *ibid.*, f. 278. MAE, CPC Turquie 147, Veillet-Dufrêche (Administrative Division) to Political Division, 2/08/1909, f. 175–9.
57 N. Yïldïrïm, Tanzimat'tan Cumhuriyet'e Koruyucu Saglik Uygulamalarï, *Tanzimat'an Cumhuriyet'e Ansiklopedisi* (Istanbul: Ilitisim Yayïnlarï, 1985), p. 1326.
58 MAE, CCC Djedda, consul (Dubreuil), 15 May 1868 and 15 February 1869.
59 MAE CCC Djeddah (Lavalette de Monbrun), 27 August 1856.
60 Proust, *Essai sur l'hygiène internationale*, p. 2.
61 S.-A. Fauvel, *Rapport sur les mesures à prendre en Orient pour prévenir de nouvelles invasions du choléra en Europe*, annex to the minutes nº 29 (28 May) of the Internationale Sanitary Conference, August 1866 (Constantinople: Imprimerie du Levant Herald, 1866), pp. 9–14.
62 Quoted by A. Siegfried, *Vue générale de la Méditerranée* (Paris: Gallimard, 1943), p. 171.
63 *Ibid.*, p. 177.
64 Procès-verbaux de la Conférence sanitaire internationale de Paris (1912), MAE, Fol LF 128–99 (1), f. 122.
65 MAE Turquie 147, Administrative Division to Bompard, 23 June 1909.
66 The three main subjects of international health protection according to Professor Santoliquido: Duguet, *le Pèlerinage de la Mecque*, p. 192.
67 M. Weber, *Économie et société* (Paris: Plon, 1971), part III, sections 6–8.
68 Fukuyama, *State-building*, p. 134.

69 P. Zylberman, 'Mosquitos and the *komitadjis*: Malaria and borders in Macedonia (1919–1938)' in I. Borowy & W.D. Gruner (eds), *Facing Illness in Troubled Times: Health in Europe in the interwar years* (Bern: Peter Lang, 2005), pp. 305–43.

70 A. Stampar, 'On health politics, 1919' in A. Stampar, *Serving the cause of public health: Selected papers of Andrija Stampar* edited by M. Grmek (Zagreb: Andrija Stampar School of Public Health, Medical Faculty, University of Zagreb, 1966), p. 78.

71 B. Konstantinovic, *La Yougoslavie pour la santé publique* (Belgrade: Institut Balkanique, 1937), p. 44.

72 C.E.A. Winslow, 'Malaria control in Italy, Albania and Macedonia', RAC/RG2/stack 1929/554/3729, pp. 8–9, 15; L. Killen, 'The Rockefeller Foundation in the first Yugoslavia', *East European Quarterly*, 34 (1990): 356.

73 N. Ferguson, *The cash nexus: Money and power in the modern world, 1700–2000* (London: Allen Lane & Penguin Press, 2001), p. 374.

74 M. Ignatieff, *Empire Lite: Nation-building in Bosnia, Kosovo and Afghanistan* (London, 2003), p. 2, quoted by Ferguson, *Colossus*, p. 165.

75 A.M. Moulin, *Le dernier langage de la médecine: histoire de l'immunologie de Pasteur au sida* (Paris: Presses Universitaires de France, 1991), pp. 244–57.

76 J. Ancel, *Géographie des frontières* (Paris: Gallimard, 1938), p. 185, referring to F. Haushofer, *Grenzen in ihrer geographischen und politischen Bedeutung* (Berlin: Wowinckel, 1927); and D. Diner, *Beyond the conceivable: Studies on Germany, Nazism and the Holocaust* (Berkeley: University of California Press, 2000), pp. 26–48.

77 D.F. Musto, 'Quarantine and the problem of AIDS', *The Milbank Quarterly*, 64 (1986): 109.

78 Konstantinovic, *La Yougoslavie pour la santé publique*, p. 59.

79 H. Zinsser, *Rats, lice and history* (London: Penguin Books, 2000 [Routledge, 1935]), pp. 133, 213.

80 S. Adamets, *Guerre civile et famine en Russie: Le pouvoir bolchevique et la population face à la catastrophe démographique entre 1917 et 1923* (Paris: Éditions de l'Institut d'Études Slaves, 2003), pp. 187, 221, 231; P. Gatrell, *A Whole Empire Walking: Refugees in Russia during World War I* (Bloomington: Indiana University Press, 1999).

81 Résolution adoptée par la Conférence de Varsovie le 22 mars 1922, appendix no. II [LN C.152.1922–C.I.E. 24], MAE Europe 1918–1940 Pologne 272, fol. 206. These contemporary estimates have been increased through present-day research.

82 M. Balinska, 'Assistance and not mere relief: The Epidemic Commission of the League of Nations, 1920–23' in P. Weindling (ed.), *International health organisations and movements, 1918–1939* (Cambridge: Cambridge University Press, 1995), pp. 84, 89.

83 Résolution adoptée par la Conférence de Varsovie le 22 mars 1922, appendix no. II.

84 Balinska, 'Assistance and not mere relief', pp. 81, 93 et 96. The data gathered by M. Kacprzak, *l'Hygiène publique en Pologne* (Warsaw, 1933), LN, CH/EPS/164, p. 124, are very different: the number of cases in Poland was at a maximum in 1919 (219,088) and decreased steadily thereafter (42,724 in 1922), whereas the number of deaths reached its climax in 1920 (22,575) then fell below 5,000 in 1922.

85 Kacprzak, *l'Hygiène publique en Pologne*, pp. 116–17.

86 W. Rose, 'Conditions in eastern European Countries', 14 March 1920, Rockefeller Foundation Archive RAC/1.1./700 Europe/16/121.
87 L. Rajchman to W. Rose, 18 February 1922, RAC/1.1./100 International/20/165.
88 Rose, 'Conditions in eastern European Countries'; L. Rajchman to L. Bernard, 27 February 1922, LN R820/12B/26208/11346.
89 Balinska, 'Assistance not mere relief', pp. 81–108; P. Weindling, *Epidemics and genocide in eastern Europe, 1890–1945* (Oxford: Oxford University Press, 2000), pp. 140–8, 163–71.
90 L. Rajchman, 'Records of the European conference held at Warsaw from March twentieth to 28th 1922', LN C.144.M.96.1922, R835/12B/19914/18972; and League of Nations, 'Conférence sanitaire européenne réunie à Varsovie du 20 au 28 mars 1922, Genève, 3/4/1922', MAE, Europe 1918–40/Pologne 272/221.
91 *Ibid.*
92 On political geography, see M. Korinman, *Quand l'Allemagne pensait le monde. Grandeur et décadence d'une géopolitique* (Paris: Fayard, 1990), pp. 132, 139ff.; D.T. Murphy, *The Heroic Earth: Geopolitical thought in Weimar Germany, 1918–1933* (Kent, OH: Kent State University Press, 1997), pp. 30–2.
93 Geouffre de Lapradelle, *La frontière*, p. 226.
94 O. Lattimore, *Studies in frontier history* (Paris: Mouton, 1962), p. 113.
95 *Ibid.*, p. 470.
96 M. Albert & L. Brock, 'Debordering the world of states: New spaces in international relations', *New Political Science*, 35 (1996): 70.
97 But see the qualifications made by V. Shkolnikov, F. Meslé and J. Vallin, 'La crise sanitaire en Russie: I. Tendances récentes de l'espérance de vie et des causes de décès de 1970 à 1993', *Population*, 4–5 (1995): 907–44; and 'La crise sanitaire en Russie: II. Évolution des causes de décès: Comparaison avec la France et l'Angleterre (1970–1993)', *Population*, 4–5 (1995): 945–82.
98 J. Ancel, *Géopolitique* (Paris: Delagrave, 1938), p. 90.
99 C. Rosenberg, 'What is an epidemic? AIDS in historical perspective', *Daedalus*, 118 (1989): 2–3.
100 Dr Mark Ryan (WHO), interviewed by K. Bradsher & L.K. Altmann: 'SARS is subsiding, but as unpredictably as it surfaced' in the *International Herald Tribune* (23 July 2003).
101 Proust, *la Défense de l'Europe*, pp. 285–92; Dr Catelan, 'Rapport sur le choléra au Hidjaz en 1890', *Archives de médecine navale*, 59 (1893): 51–4.
102 W.R. Roff, 'Sanitation and security: The imperial powers and the nineteenth-century Hajj' in R.B. Serjeant and R.L. Bidwell (eds), *Arabian Studies*, VI (London: Scorpion Communication, 1982), pp. 143–60.
103 Gaddis, *Surprise, security, and the American experience*, p. 86; P. Baldwin, *Disease and Democracy: The Industrialized World faces AIDS* (Berkeley: University of California Press, 2005), pp. 1–6, 227–43, explores in depth this 'fundamental continuity with past tactics' in the management of epidemics.
104 L. Murard & P. Zylberman, 'The Health Organization of the League of Nations: Functionalism revisited', unpublished paper submitted to the conference *Health in the city: A history of public health* organized by the Society for the Social History of Medicine and the International Network for the History of Public Health, Liverpool 4–7 September 1997.
105 L. Garrett, 'The politics and profits of AIDS', *International Herald Tribune* (17–18 July 2004).

3

Yellow Fever Crusade: US Colonialism, Tropical Medicine, and the International Politics of Mosquito Control, 1900–1920

Alexandra Minna Stern

From 1900 to the 1940s, the United States spearheaded an international campaign against yellow fever, channeling massive resources and sanitary know-how into the eradication of the mosquito vectors of this viral disease, their larvae, and the standing water in which they copiously bred. For the most part, this crusade unfolded in Latin America and the Caribbean and, as such, was an integral facet of expanding US scientific and cultural hemispheric dominance in the Americas. In 1900, US physicians conducted a series of experiments in Cuba that conclusively demonstrated that yellow fever was not transmitted via the air nor through infected bedding and clothing but via the female *Aedes aegypti* (initially called the *Stegomyia fasciata*) mosquito. These findings furnished the basis for a top-down mosquito control effort in Havana, a city occupied by US military and sanitary forces in the wake of the 1898 Spanish-American War. The successes of this operation, which reduced yellow fever deaths to zero in under one year, were transferred to the Panama Canal in 1904, when the United States gained possession of this Central American isthmus in order to build a trans-oceanic waterway. And after the Canal was opened in 1914, the United States was sufficiently concerned about the threat of disruptive diseases spreading out from this new geographical opening to embark on a major yellow fever eradication campaign that lasted for over 20 years and expended nearly 6 million dollars.[1]

The story of the yellow fever crusade reveals important aspects of the emergence of key paradigms, patterns, and problems of modern global health. First, this story shows the extent to which US health entities and professionals, belonging both to military and federal agencies and to philanthropic institutions, influenced the development of tropical medicine through unwavering attention to yellow fever. Because US colonial ventures, particularly those entailing the occupation and territorial possession of foreign lands, have often been regarded as exceptional or anomalous, the connections between US colonialism and tropical medicine has been underappreciated by scholars. This relationship has been amply overshadowed by accounts of British, and to a lesser degree, French tropical medicine in colonized locales such as India and Egypt. This historiographic lacuna is even more pronounced in the case of Panama, a long-colonized republic which usually appears as an afterthought in critical studies of early-twentieth US colonialism and overseas territorial expansion.

Second, this story elucidates a distinct cartography of colonial medicine, one that pivots on the Panama Canal and stretches into the Caribbean and down both the tropical Atlantic and Pacific coasts of South America. More specifically, this yellow fever map illustrates how US commercial interests, anxieties over migrating microbes, presumptions about racial backwardness, a medico-military brand of disease containment and control, and at times, a humanitarian desire to institute basic public health services in less industrialized and poorer countries, merged in a dynamic, uneasy, and asymmetrical fashion in the first half of the twentieth century. In waging their war against yellow fever and mosquitoes, US health authorities mentally mapped a medicalized region that was at once rife with danger and the potential of redemption. In this sense, US tropical medicine was critical to the imagining of 'Latin America' as an object of analysis and geographical entity in need of expert intervention and scientific management.[2]

Third and finally, even as this story underscores the predominance of medical reductionism in approaches to yellow fever control and mosquito eradication, it repeatedly illustrates the limits of such narrow health strategies and exposes the messier compromises made on the ground by the heterogeneous actors involved in the implementation of sanitary policies. An examination of the quotidian practices of mosquito control, which in the case of yellow fever targeted the *Aedes aegypti* (or *Stegomyia fasciata*), a domestic species that never strayed far from human habitation, can offer insights into the social construction and contestation of the boundaries of public and private, clean and infected, and modern and primitive in the early twentieth century. Furthermore, unlike diseases such as hookworm and typhoid, whose causative microorganisms could be identified in the laboratory from specimens of bodily fluid or excreta and treated prophylactically or therapeutically, the campaign against yellow fever was waged against the annoying and insidious mosquito vector and, until the 1930s, an enigmatic virus unseen by the pre-electron microscopes of the era.

This chapter explores the yellow fever crusade launched by the United States and focuses on the formative period from 1900 to 1920. Over these two decades, the task of eradication shifted from the purview of the US military, as embodied by the army medical corps in Cuba and Panama, to US philanthropy, in the form of the Rockefeller Foundation's Yellow Fever Commission. Against the backdrop of this chronology, I trace the complex interplay between US colonialism, tropical medicine, and yellow fever control during a historical moment in which optimism about the promise of total sanitary control and disease eradication was on an unshakeable and imperious ascendance.

Experimenting with yellow fever in Cuba

In the late nineteenth century, as European nations embarked on colonial endeavors in Africa and Asia, the United States too entered the game. With the rationale of ousting Spain from its remaining colonies in the Caribbean and Pacific and concomitantly supporting the independence struggles of indigenous 'freedom fighters', in 1898 the United States started the Spanish-American War of 1898, eventually taking control of Cuba, the Philippines, and Puerto Rico (the latter remains a US colony to this day).

This surge of European and US colonialism was intimately linked to the birth and consolidation of tropical medicine.[3] On one level, developments in bacteriology, germ theory, parasitology, and entomology coincided with late nineteenth-century colonial and imperial projects. Yet these developments often relied heavily on the laboratories set up by colonial physicians in places like Formosa, India, and West Africa, and on research experiments with colonized subjects. If the aim of tropical medicine was to identify and eliminate novel classes of diseases transmitted in or through parasites, insects, and helminthes, the paramount concern of its practitioners was usually not the well-being of native populations but sanitation as the first step to the sustained settlement of white Europeans and Americans. In medical parlance, this meant that the focus was on the protection of the non-immunes, who came to the tropics biologically unprepared for the results of an insect bite. Unlike many native residents of colonial zones, who had grown up surrounded by local disease ecologies, colonial newcomers had no acquired immunity or resistance. The great irony, of course, is that colonization itself – the relocation of thousands of non-immunes en masse to a tropical locale and any attendant infrastructural and ecological disruptions – produced a ripe disease reservoir that fostered the conditions for severe infectious outbreaks.[4]

Not surprisingly, when hundreds of US soldiers landed in Cuba in 1898, yellow fever, which had occurred periodically on the island since the eighteenth century, quickly erupted, killing more men than had military combat. In order to quell this situation, the United States turned to two army doctors: Walter C. Reed to form a commission to determine the transmission patterns of yellow fever; and William C. Gorgas to institute sanitary control in Havana and its environs.

For Gorgas, fighting yellow fever and its winged transmitters necessitated a major shift in medical thinking that was, in turn, catalyzed by the results of Reed's experiments. Gorgas was a career army doctor originally from the South who had spent the bulk of the 1880s and 1890s in the American West (Texas and South Dakota), where he simultaneously pursued frontier medicine and territorial administration.[5] In 1897 he was sent to Fort Barrancas, Florida, to contain a yellow fever outbreak. In accordance with the conventional medical wisdom of the day, Gorgas proceeded to carefully isolate the sick, fumigate known sites of infection, and disinfect all potentially contaminated clothing, bedding, and objects. When Gorgas was ordered to Havana the following year to administer to the troops, he was expected to carry out the same measures in order to contain yellow fever among US soldiers.[6]

When he arrived in Havana, Gorgas went about 'cleaning up' the city in good sanitarian fashion: he purified the water supply, improved the sewer system, scoured the streets, and disinfected known sites of contagion.[7] These efforts, however, did nothing to stave off yellow fever, a virus that had struck the Americas in epidemics over the past centuries, including a severe 1878 outbreak in the Mississippi River Valley, because mosquitoes did not care if the streets were pristine or the bedding steam cleaned, they just coveted a container of standing water, no matter how humble or obscure, in which to propagate.[8]

As Gorgas sought to understand why his methods of water purification and broad-based sterilization, which had worked well against typhoid, were futile against yellow fever, his superior Reed was at a military camp down the road carrying out an experiment that he hoped would demonstrate that yellow fever was transmitted by the *Aedes* and the *Aedes* alone.[9] In large part, Reed was simply attempting to systematically prove, with the benefit of a sizable subject population of ready volunteers, the prescient hypothesis of the Cuban physician Carlos Finlay. Since 1881, Finlay had maintained in the face of a great deal of scientific ridicule that a mosquito, more precisely the female *Stegomyia*, was the vector responsible for yellow fever transmission. Finlay, however, had failed to reproduce yellow fever infection in humans with the *Stegomyia* because his calculations about the timing of infectivity were incorrect. Relying on Finlay's insights and on the Cuban's prized mosquito collection, the Reed Commission (which included Dr James Carroll, Dr James Lazear, Finlay and his compatriots Juan Guiteras and Aristides Agramonte) established that yellow fever was indeed spread by one mosquito species (the *Stegomyia* or *Aedes*), never through fomites (bed sheets, clothes, or other inanimate objects). They demonstrated that it was submicrosopic and filterable and followed particular patterns of gestation in the mosquito and incubation in the infected human.[10] Furthermore, Reed and his colleagues, such as the discerning Dr Henry Rose Carter, showed that humans were capable of spreading viral infection via the mosquito vector 1 to 3 days after infection and that the yellow fever microorganism required 10 to 15 days to develop in the *Aedes* before it could be transmitted to another organism. Notwithstanding several deaths among his research subjects (including Lazear, soon eulogized as a martyr of modern science), Reed and his colleagues confirmed and clarified Finlay's hypothesis. Eager to receive credit for their impressive findings, Reed and his collaborators swiftly presented and published their results, which Gorgas, in regular contact with Reed, watched unfold in real time.

At the same time that the Reed Commission was immersed in its experiments, the concept of 'tropical disease' and the medical field of tropical medicine were cohering, due largely to the work of the British physician Patrick Manson, who discovered filariasis, and Ronald Ross, whose focus was malaria. Between 1898 and 1902, Manson defined tropical diseases as a novel class of principally vector-borne parasitic infections and founded the London School of Tropical Medicine. It was Manson who mentored Ross, who was based in India, and had spent much of his life traveling in colonial circuits. Manson guided his student in the quest to identify *plasmodium* in the guts of female *Anopheles* and demonstrate through the methods of germ theory that this microorganism was solely responsible for malaria.[11] Understandably, these two men and the diseases they studied have become central to the story of the development of tropical medicine. To some extent, their enshrining as the progenitors of tropical medicine has obscured other historical medical actors including Reed, Gorgas, and certainly their Cuban counterparts. While the Reed Commission did not identify and name the microbe responsible for yellow fever, it did establish the disease's transmission patterns and confirmed the role of the mosquito vector.

Taken together, these discoveries and experiments helped to transform the mosquito from an annoying pest into a deadly predator. Moreover, they indicated the parameters for new strategies of disease containment and control. By 1900, Gorgas, who 2 years earlier had been highly skeptical of the mosquito-vector theory, was convinced of the need for another type of sanitary campaign. Given that the *Aedes* was fond of lurking in flower pots, cisterns, containers with rain water, and other spots of standing water (unlike the *Anopheles*, a wilder, more wide-ranging creature), Gorgas elected to institute a totalizing system of mosquito extermination that divided Havana into sanitary districts. Each district was overseen by a medical team that kept a detailed inventory of file cards on every house and water source.[12] After this data was compiled, Gorgas's brigades drained, oiled, or capped all wells, cisterns, and ponds, and fumigated homes in order to kill adult mosquitoes. This strategy worked. Within 3 months yellow fever had decreased markedly, from 1,400 cases in 1900 to 37 in 1901, and within a year had all but vanished.[13] Keeping abreast of Gorgas's work from the East

Coast, Reed repeatedly commended his colleague, telling him in June 1901, 'I am astonished at the strength of your Mosquito-Destroying Sanitary Squad'.[14]

Through the yellow fever experiments in Havana, and as the United States was on the brink of acquiring the Panamanian isthmus, the emergent field of tropical medicine was transforming techniques of, and ideas about, disease etiology and eradication. Theories of filth, miasma, and putrefaction were being supplanted by theories of microorganisms, frequently visible under the microscope, that were responsible for diseases such as anthrax, rabies, cholera, and typhoid. Moreover, scientists were starting to identify a hitherto unrecognized universe of mobile disease vectors, insects and rodents, which unwittingly transmitted germs across various species lines to humans and often from human to human. As Nancy Stepan has shown, this process of magnification helped to generate a new visual iconography of the tropics as replete with disease-carrying organisms that only tropical specialists could identify and had the expertise to eliminate.[15] Indeed, tropical medicine was integral to a re-imagining of the tropics from a 'white man's grave' into a tamable paradise. Over and over again, in writings from the turn of the century, American and European journalists, scientists, and commentators praised the discoveries of Manson, Ross, Reed, and others, who had confirmed that the tropics were not only livable for whites but wide open for total colonization. As Gorgas told the graduating medical class of Johns Hopkins University just a year before the Panama Canal was completed, 'our sanitary work at Panama will be remembered as the event which demonstrated to the white man that he could live in perfectly good health in the tropics; that from this period will be dated the beginning of the great white civilization in these regions'.[16]

Eradicating yellow fever in Panama

If Cuba was the starting point for a convergence between US colonialism and tropical medicine, Panama soon became its epicenter. Propelled by many factors, above all an interest in enlarging the US navy and a push for commercial expansion, in the late nineteenth century the United States reignited a decades-old plan to forge a transoceanic passage between the Atlantic and Pacific through Central America. In the 1880s, a similar attempt by the French had failed miserably, culminating in disaster and political scandal. The French had been stymied by many factors but foremost among them was the relentless outbreak of disease, particularly of mosquito-borne ailments, which ultimately took the lives of approximately 20,000 able-bodied workers.[17]

Continuing a pattern of imperial expansion, President Theodore Roosevelt orchestrated the secession of Panama (upper Colombia) from Colombia and the Hay-Bunau-Varilla Treaty was signed in 1903, placing jurisdiction of the

Canal Zone (about 50 miles long by ten miles wide, or 500 square miles total) in US hands. Technically and legally, the Canal Zone was an 'unorganized possession', similar to Guam and American Samoa, not an 'incorporated territory' like Alaska and Hawaii nor an 'unincorporated territory' like the Philippines and Puerto Rico. It was accorded virtually no political autonomy until the mid-century and was not fully transferred back to the republic of Panama until 1999.[18] This political arrangement extended beyond the Canal Zone proper into the two cities of Colon and Panama and informally into the interior regions of the Panamanian republic.

Once in control of the Canal Zone, the United States formed the Isthmian Canal Commission (ICC) to govern the region and oversee the gargantuan task of carving a waterway through the humid, rocky, and bug-infested riverine jungle. From 1904 to 1914, approximately 60,000 workers, West Indians, Euro-Americans, and Europeans, excavated over 230 billion cubic yards, blasted gigantic boulders, dug and dredged.[19] Their engineering blueprint was a Locks Canal, which necessitated damming much of the Chagres River near the Atlantic and the Pacific and installing a series of locks to allow ships to gradually move between sea level and the higher elevations of the Lakes.

During the decade of construction, the United States sought to manage its newly obtained 'unorganized possession' with meticulous and methodical precision, putting in place a highly paternalistic occupational and social order in which the government was responsible for everything. It controlled work schedules, lodging, food, entertainment, and the daily regimes of workers. The system, however, was strictly separated on the intersecting lines of class and race: the skilled workers were overwhelming white Americans paid in gold and the unskilled workers were overwhelming colored, mainly from the West Indies (Jamaica and Barbados, for example) and paid in silver.[20] A few black workers and a sizable minority of Basque immigrants who joined the Canal's labor force sometimes crossed this boundary. However, everything related to building the Canal and its laborers was refracted through the division of the gold and silver rolls. Building the canal required more than tens of thousands of workers, a viable blueprint, and massive financial commitment from the US government. It could not feasibly proceed without controlling the diseases that had so incapacitated railroad workers and '49ers' heading to California during the 1849 Gold Rush via Panama and foiled the French several decades later.

The Panama Canal story needs to be incorporated into our understanding of US colonialism around 1900 and the making and priorities of tropical medicine in the modern era. When the United States gained control of the Canal Zone, one of the clauses contained in the US-Panama treaty was an agreement on the part of Panama to 'comply in perpetuity with the sanitary ordinances, whether of a preventive or curative character, prescribed by the United States'. If Panama failed to do so, it relinquished to 'the US the right and authority to enforce the same'.[21] In addition, the United States was granted complete authority over sanitary matters in the cities of Panama and Colon, including the right to enter private homes and remove standing water and gutters, cut grass on private property, and fine and imprison individuals in violation of the health codes.[22] Not surprisingly, this arrangement led to many tensions, especially in Panama City, and the habitual filing of grievances against the ICC.

It was the knowledge and experience gained in Havana that Gorgas brought to Panama a few years later when he was appointed Chief Sanitary Officer. When he landed in Panama, Gorgas had few supplies – just seven assistants, including one physician and an English nurse – while the sanitary infrastructure (hospitals and machines) legally remained in French hands. He faced substantial skepticism, particularly in non-medical quarters, about the mosquito vector theory of disease transmission, and was nervous that the politicians and engineers in charge of the Canal would deny his requests and obstruct his sanitation plan. And, in fact, Gorgas did initially encounter opposition from the first head of the ICC, a situation exposed in a scathing 1905 report on sanitation in the Canal Zone commissioned by the US Secretary of War.

Nevertheless, due to formidable pressure from President Roosevelt, by late 1905, the Chief Engineer of the ICC at the time, John F. Stevens, authorized Gorgas to employ a full team in Panama and granted him $50,000 to launch his campaign.[23] Gorgas quickly swung into action. Following his Havana model, he designated 25 sanitary districts, each assigned an inspector with a team of 20 to 100 men.[24] Dozens of sanitary workers started drainage projects, carried out house-to-house inspections and fumigation, constructed mosquito covering and netting, applied agents (kerosene, sulfur, and alcohol) to destroy mosquitoes and larvae, and dug, cleared, and lined ditches and water channels. US physicians took over the Ancon and Colon hospitals, admitting hundreds of patients.

According to Gorgas, 1905 and 1906 were the height of the sanitary campaign, especially with respect to the first item on his agenda, yellow fever eradication: 'in looking back over our ten years of work, these two years of 1905 and 1906 seem the halcyon days for the Sanitary Department. It was really during this period that our work was accomplished'.[25] By May 1906, for example, Gorgas could report the complete eradication of yellow fever, a fact he proudly reiterated in his monthly reports.[26] During January 1906, a typical month during these 2 years, nearly 10,000 houses in Panama City were inspected by the *Stegomyia* brigade, which found and destroyed mosquito larvae in 1,785 of those sites.[27] These statistics were repeated over and over again. Because of the different propagation patterns of the *Anopheles* and *Stegomyia*, the yellow fever teams scoured the urban areas while the malaria brigades concentrated on rural areas. In 1906, for example, in and around Panama City alone, the *Stegomyia* Brigade carried out a grand total of 117,319 home inspections, 2,160 reinspections, and found and oiled larvae in 11,043 houses. The *Anopheles* Brigade cut and burned 681,300 square yards of grass, dug 5,903 linear feet of ditches, used 169 containers of mosquito oil, laid 1,950 linear feet of cement, and filled 244 wells.[28] After the last case of yellow fever was reported in late 1905, Gorgas continued to monitor the yellow fever situation, keeping tabs on the mosquito index, and continuing to send out sanitary teams that searched neighborhoods and homes for any existing or potential source of standing water in which *Aedes* could breed.[29]

Over this period, approaches to mosquito control changed, and Panama is especially illuminating in this regard. During the first few years of Gorgas's mosquito extermination campaigns, first in Havana and then in Panama, his strategy concentrated, in the following order on: 1) preventing mosquitoes from biting people with screening and isolation; 2) killing mosquitoes that might have become infected by burning sulfur or pyrethrum; 3) instituting quarantines on yellow fever areas so that potentially infected people could not travel to areas with uninfected people; and 4) organizing brigades for the destruction of mosquito larvae.[30] The dominant logic here was primarily one of quarantine, centered first on controlling the infected human, then on the adult mosquito, and finally on the larvae. This order of implementation reveals the extent to which a nineteenth-century mindset of quarantine influenced initial yellow fever campaigns, even as scientists recognized that vector-borne diseases demanded a distinct repertoire of strategies. Within one decade, this order was reversed, as the primary aim became eradication of all potential spots of mosquito breeding and the destruction of larvae. Ross declared that the foremost motto of tropical sanitation should be 'No Stagnant Water'. And writing in 1909, Rose Carter, a United States Public Health Service surgeon who had served with Gorgas in both Cuba and Panama and was instrumental in discovering the intrinsic incubation period of yellow fever in the *Stegomyia*, stated '*We could not control the human host* to the degree necessary for success'.[31] Simply put, controlling the insect side of the infection chain was much easier and by 1906 had become the cornerstone of mosquito campaigns.

Even as the focus on the mosquito shifted the emphasis away from people onto insects, mosquito eradication campaigns often entailed a heavy-handed intrusion into homes and private domains. Just as the *Aedes* was classified as a domestic mosquito, yellow fever control became a thoroughly domestic affair subject to regulation by Gorgas and his medical corps, an arrangement that led to tensions in many instances. For example, based on his authority to control sanitary matters in Panama City and Colon, Gorgas insisted that all gutters be removed from houses; those that did not comply within the mandated period were fined, and if they failed to pay, jailed. More broadly, when met with resistance or conflict, US sanitary interventions in Panama usually resulted not in the protection of the invaded and his or her home, but in the rearticulation of the supreme authority of the United States in all health matters. For example, in 1913, Jorge Domingo Arias, a resident of Panama City, filed a complaint against ICC sanitary inspector, J.M. Carpprow, for destroying five kitchens on his property (probably rental units) that were deemed to be an unsanitary nuisance. After many letters were sent back and forth on his behalf from Panamanian to US officials, the final outcome was an official reiteration of Article VII of the 1903 treaty, namely that: 'the Republic of Panama agrees that the cities of Panama and Colon shall comply in perpetuity with the sanitary ordinances whether of a preventive or curative character, prescribed by the United States and in case the Government of Panama is unable or fails in its duty to enforce this compliance'.[32]

The flipside of intrusion, however, was neglect. In a somewhat counterintuitive fashion, the shift in focus from the human to the mosquito provided a justification for racial segregation in Panama and the colonial tropics in general. The formation and implementation of tropical medicine bolstered the hubris of the Americans, who attributed their newly found vigor and mobility to their resilient racial make-up and civilized manners, especially in contrast to the unhygienic customs of 'primitive' peoples lower down on the evolutionary ladder.[33] Once combined with the logic of protecting the non-immunes, usually categorized as white colonials or gold-roll employees, this translated into calls for segregated living quarters, so that 'native' reservoirs of disease would be geographically distant from the colonials' enclaves. As Carter explained, 'To prevent the infection of the mosquitoes which have access to the men we are protecting is simply to segregate the quarters of these men from those of the natives and colored laborers – a source of infection to the insects – a sufficient distance,' something that 'has been inculcated by the British writers for years'.[34] This tenet was also central to Ross, who wrote in his 1902 manual *Mosquito Brigades and How to Organize Them* that 'the good health enjoyed by the British in most Indian stations is probably largely due to the fact that they live apart from the natives in separate cantonments'. Ross encouraged physicians to hunt down mosquitoes by placing 'natives', who were presumed – often incorrectly – to be immune in non-netted quarters, and waiting for mosquitoes to enter and seek out human skin.[35]

Several dispatches written by United States Public Health Service physician James C. Perry, who was one of the first health officers stationed in Panama, captures the animus of US health officers. Describing Colon's urban districts, Perry evinced dismay with the conditions of the city's poorer neighborhood, which extended over a swampy area: 'These habitations are filthy in the extreme, and it is difficult to understand how people can live in such unsanitary surroundings with any semblance of health'.[36] Like many of his contemporaries, Perry often saw cleanliness and dirt, health and disease as reflections of racial differences in personal and public hygiene, wherein dark-skinned inhabitants were perceived to be dirtier and more unkempt than their fairer counterparts. Such prejudices constituted the ugly underbelly of tropical medicine, which, at least until the 1940s, and many would argue that in many ways even to this day, partook of the white man's burden to lift up, civilize, and sanitize the natives. For Perry, this mission involved making Colon into a 'healthy tropical city' through the two-pronged approach of sanitary engineering and hygienic education.[37]

In the Panama Canal, the apparent enhanced health of whites and the diminished health of blacks rapidly became a self-fulfilling prophecy since it was the quarters and districts of the whites, engineers and laborers, where mosquitoes were most vigorously attacked. In juxtaposition to their white counterparts, the homes of West Indian workers were not systematically screened and the standing water in their communities less likely to receive immediate attention. The lengthy monthly and annual reports of the ICC Health Board reveal the vast disparities in the screening of the living quarters of gold and silver employees, a situation that undoubtedly contributed to ongoing albeit declining numbers of malaria cases among West Indians in the segregated work towns stretching along the Canal from Panama to Colon. Again, the purported non-immunity was often the rationale for extensive mosquito brigades in areas with a high density of colonials or white employees.

In the Panama Canal, this segregation was in part a reflection of Jim Crow racism, as demonstrated in the silver and gold rolls – in housing, schools, civic organizations, and on the hospital wards. However, medicalized ideas about racial differentiation were also being produced in Panama itself where West Indian and colored patients were repeatedly used as human subjects for clinical studies, many of which supported the thesis that blacks and whites were constitutionally different. For example, after analyzing 500 cases of syphilis among black males in the Ancon Hospital over a period of 23 months (from 1911 to 1913), Dr Walter G. Baetz, a member of the Medical Association of the Isthmian Canal Zone, concluded that 'certain syphilitic manifestations' were much more prevalent in 'tropical negroes than among Caucasians in their northern abodes'.[38] Interestingly, he asserted that blacks were much more likely to suffer from glandular, usually genital enlargement, whereas nervous system involvement was more common among whites, a finding that could easily perpetuate underlying associations of blacks with abnormal sexuality and whites with troublesome disorders, such as neurasthenia, that resulted from degeneration and over-civilization.

If black West Indians were visible as subjects of many of the clinical studies conducted in the Ancon and Colon hospitals, their lives and labor were invisible or marginalized in almost every other way. They constituted the majority of the workers that physically built the canal and carried out the sanitation campaigns – in 1907, for example, there were almost 11,000 white versus almost 29,000 black employees. Yet the latter were repeatedly marginalized or erased from the official documents. For instance, buried in the appendices of President Roosevelt's 1906 otherwise sanguine report on the glories of Panama and the wonderful results of the health crusade, were mortality figures which actually revealed that while white workers benefited from the health regime blacks suffered. From January to October 1906, for example, 17 whites died per 1,000, as opposed to 59 per 1,000 blacks.[39] Despite an occasional acknowledgment that the dismal health condition of blacks was due to a lack of sanitation, Gorgas performed the same acts of erasure in his monthly and annual reports. For example, Gorgas *only* included the names, occupations, and ages of deceased gold roll- or white employees, their black counterparts were reduced to mere statistics. While Gorgas did acknowledge that the death rates of 'Negro employees' were substantially higher than that of whites (13 versus 4 per cent, for instance, in 1907), he was consistently more interested in documenting the superb health of whites even when comparing the Canal Zone to an American town or city. He boasted in a 1909 dispatch: 'I call attention to the very small rate among our 5,179 white employees from the United States, among whom we had 1 death, giving us an annual death rate of 2.31 a thousand', adding that it 'would be very hard to find a community of the same size anywhere at home having as small a death rate as 2.86 per thousand'.[40]

Clearly, the victims of the success of narrow and targeted campaigns against specific pathogens (such as yellow fever and malaria) were the silver workers, usually black West Indians, who received substandard health care and suffered from significantly higher mortality and morbidity rates, above all from diseases, such as pneumonia and dysentery, which fell outside the purview of the tropical medicine showcase. From 1904 to 1910, for example, deaths from pneumonia constituted at least 25 per cent of the total deaths (including accidents) and were suffered by silver-roll workers primarily hailing from Jamaica, Barbados, and Colombia. During the 1910s, these strategies of racialized segregation traveled with many of the physicians who had spent time in Panama back to the continental United States and were implemented in sites such as Chinatowns on the West Coast and along the US-Mexico border.[41] The most devastating irony is that, literally, the physical burden of disease eradication, which largely benefited whites and their families; the spraying, oiling, and draining, was carried on the backs of West Indian and black laborers.

'World-wide eradication': the Yellow Fever Commission

Although yellow fever eradication in Panama was effective as a narrowly targeted measure, it was carried out in a public health vacuum. Instead of working to bolster one facet of a larger medical infrastructure, yellow fever control concentrated exclusively on the elimination of standing water and the achievement of a low-mosquito index (the ratio of containers with larvae found as compared to the number of sites examined, usually calculated at 2 per cent).[42] The limits of this approach were clear to the Rockefeller Foundation (RF), which arrived in Panama in 1914 to launch a hookworm program and found uncinariasis infection rates of 80 per cent or higher in many provincial areas.[43] Not only had Gorgas and the ICC's overwhelming interest in yellow fever and malaria stymied the control and treatment of much more mundane but just as deadly diseases such as dysentery and pneumonia, but the US stranglehold on the health system had held back the growth of the Panamanian health system. As the annual report of the RF's International Health Board stated in 1921, 'the development of local initiative' in Panama 'has been stifled by the paternalistic policy of the Canal Zone'.[44]

Despite evidence that the sanitary strategies employed by the United States in the Panama Canal might be problematic, the RF decided to apply the lessons of yellow fever control learned in Panama throughout the Americas. The yellow fever paradigm, which started in Cuba and expanded in Panama, was just too seductive and successful in the eyes of US health authorities to challenge or revise. Moreover, rising anxiety about the potential that yellow fever might re-emerge once ships began to cross the Panama Canal, and travel either up to the United States or across the oceans to Asia or Europe, prompted the RF to establish the Yellow Fever Commission (YFC) in 1916. Because of his proven track record and international stature, the RF chose Gorgas to head the commission. The YFC's objective was to identify the endemic foci or seed-beds of yellow fever infection in the Americas, which were primarily port cities that constituted key nodes in the chain of inter-American maritime commerce. The composition of the commission, which brought together physicians who had collaborated in Cuba in 1900, including Gorgas, Guiteras, and Carter, as well as three other scientists, reflected the extent to which US colonial medicine flowed rather seamlessly from military and uniformed agencies into the RF's philanthropic mission. Clearly, the RF was emboldened by the prospects of total sanitary conquest and continued American medical dominion: 'to eradicate yellow fever from these seed-beds and thus to rid the world of the disease, is the high adventure upon which the Rockefeller Foundation embarked in 1918 under the leadership of General Gorgas'.[45]

After an initial fact-finding mission in the summer of 1916, the YFC identified one seed-bed of yellow fever: Guayaquil, Ecuador. After waiting for World War I to end so that Gorgas could lead the campaign, the YFC set out to eradicate *Aedes* from Guayaquil, at the same time that it continued to monitor the Brazilian eastern coast and the Caribbean for signs of the mosquito-borne disease and plotted the extension of its field investigations to Mexico and West Africa. The eradication method replicated what had been done first in Havana and in a much more comprehensive and prolonged fashion in Panama. Guayaquil was divided into districts, each supplied with a sanitary squad. Standing water was eliminated, larvae destroyed, and within six weeks yellow fever had been contained.[46]

One of the most striking aspects of the YFC's work in Guayaquil was the conflict generated over the need for a revamped water and sewage system in the city, which several YFC members deemed necessary to rid the city of multitudinous spots of mosquito breeding places. Given its infatuation with the narrow eradication approach, the International Health Board Director, Wycliffe Rose, refused to invest in what he considered to be the responsibility of the Ecuadorian government. As John Farley has argued in the case of Guayaquil, 'the Health Board's goal, as Rose made clear in his Working Plan for Ecuador, was to eradicate yellow fever from Guayaquil in order to eliminate it from the west coast of South America. They could not wait on the long-term betterment of water and sewage systems'.[47] The RF's unwavering faith in yellow fever eradication was illustrated by the fact that even without the cost of investing in infrastructural projects abroad, the International Health Board nevertheless managed to devote the majority of its financial resources to yellow fever, up to 50 per cent of its budget from 1925 to the late 1930s.[48]

A closer look at the YFC's initial activities from 1916 to 1920 illustrates the degree to which yellow fever eradication continued to revolve around the Panama experience. For example, during his tours of potential yellow fever hot-spots, Gorgas sent hundreds of copies of his book *Sanitation in Panama* to Latin American political and sanitary authorities, clearly viewing it as the supreme instructive manual. In addition, because the Canal was now the primary route between the Atlantic and Pacific Coasts of South America, the YFC's trips often returned to the Canal Zone to embark on the next phase of a journey, a stay-over that inevitably invoked in Gorgas many nostalgic memories of his reign on the isthmus. Finally, Panama was regularly Gorgas's point of comparison for the yellow fever eradication campaigns he and his colleagues initiated in Ecuador, Peru, and Guatemala. As he wrote in his diary in 1916, 'in case of work at Guayaquil I think it would be necessary to put up three or four properly screened and sanitary homes for our employees, such as those put up for our employees at Panama. If this were done it would not be a permanent expense as, when we were through with them, they could be sold at such a price as would cover the original expenditure'.[49] With Panama at the center, yellow fever crusaders produced a new medicalized cartography of tropical medicine in Latin America, a map that conflated racial difference, disease propensities, and vulnerable ports of entry for commerce and infectious organisms alike.

This did shift in minor ways in the 1920s, when several RF scientists helped to devise an ingenious and effective anti-larval method that broke with search-and-destroy mentality of the past. In 1918 and 1919, Dr H.H. Howard, working for the RF in Mississippi, demonstrated that certain species of fish placed in water tanks would devour mosquito larvae. While larger water tanks were more easily screened and oiled, live fish were delivered by sanitary squads to people's home tanks and replaced as needed, initiating an anti-yellow fever strategy that was tolerated if not welcomed by many Latin Americans. The product of collaboration and cross-fertilization among the US Bureau of Fisheries, the New Jersey Agricultural Experiment Station, and the Rockefeller Foundation, this simple and economical measure was implemented in Guayaquil in 1918 and resulted in a rapid decrease of yellow fever and was soon replicated in other parts of Latin America.[50]

Conclusion: The mirage of yellow fever eradication

In retrospect, the problems and pitfalls of the yellow fever crusade from 1900 to 1920 are manifold. The model of disease eradication was unholistically narrow from a public health perspective, its underlying presumptions about immunity and health based on perverted racial stereotypes about the hygienic inferiority of blacks and superiority of whites, and its implementation guided by a mixture of militaristic surveillance and intrusive paternalism. Furthermore, many of the yellow fever crusaders were myopically convinced of the rightness and righteousness of their approach. For example, until the 1930s, the Rockefeller Foundation supported the claims of its Rockefeller University researcher, the Japanese bacteriologist, Hideyo Noguchi, who asserted in 1918 that he had identified the yellow fever microorganism – which he labeled *Leptospira icteriodes* – in his temporary laboratory in Guayaquil. Although what Noguchi actually found was a spirochete bacteria not the yellow fever virus, he produced both a serum and vaccine that were administered (causing neither harm or help) to thousands of people in Latin America. It was not until the 1930s that Noguchi's findings were disproved; until then a combination of support from the powerful Rockefeller Foundation and an inter-American medical system that undervalued the research of Latin American scientists allowed Noguchi to hold sway, even as his findings were being challenged by laboratories in Brazil and West Africa.

The yellow fever campaign undertaken by US military, public health, and philanthropic organizations and actors from 1900 to 1920 constituted an important facet of US colonialism in the Americas. In regard to Cuba and Panama, yellow fever control was the banner of scientific success waved by army physicians such as Reed and Gorgas, who were entrusted with quelling disease in order to facilitate US political dominion and commercial interests. Furthermore, in the mid-1910s, as US physicians sought to institute the practices honed in Havana, the Canal Zone, and other 'hot-spots', they continued to project their racial suppositions about the hygienic ignorance of 'natives'. Without supporting the kinds of horizontal health interventions that would have improved basic infrastructural conditions, especially the water and sewer supply, US physicians helped to perpetuate the material circumstances of deprivation in which they could persistently view darker and poor inhabitants – whether West Indians in the Canal or indigenous communities in the Andes – as closer to primitiveness than modernity. This paradox was not lost on several of the RF health officers sent to Latin America to direct hookworm programs. In Panama, for instance, RF health officers spent a great deal of their time trying to build privies in remote areas and regularly faced distrust or hostility from local residents who resented the intrusiveness of their immediate predecessors, the ICC sanitary corps.

By confirming the yellow fever vector and controlling the mosquito index, first in Cuba and then in Panama, the United States earned a place at the table of early twentieth-century tropical medicine. Echoing European colonials who touted their efforts against leprosy, malaria, and trypanosomiasis (sleeping sickness), Gorgas wielded his triumph against yellow fever as evidence that civilized people, in other words Europeans and European Americans, could return safely to the fructiferous and human cradle of humanity: 'we have just reversed the process; we have just made sanitary discoveries that will enable man to return from the temperate regions to which he was forced to migrate long ages ago, and again live and develop in his natural home, the tropics'.[51] If such proclamations reflected the exuberant arrogance expressed by many medical men during tropical medicine's colonial heyday, a little over a decade later they would begin to ring hollow. By the 1930s, cases of yellow fever were appearing in the Amazon, an epidemiological phenomenon that revealed that species of mosquito other than the *Aedes* could transmit yellow fever.[52] Increasingly, for many RF officials, these setbacks signified that yellow fever eradication had irrevocably failed, particularly when judged against earlier lofty goals of world-wide eradication.[53]

Notes

1 M. Cueto, 'The Cycles of Eradication: the Rockefeller Foundation and Latin American Public Health, 1918–1940', in P. Weindling (ed.) *International health organizations and movements, 1918–1939* (Cambridge: Cambridge University Press, 1995), pp. 222–43.

2 For a superb discussion of this phenomenon see R. Salvatore, 'The Enterprise of Knowledge: Representational Machines of Informal Empire', in G.M. Joseph, C.C. LeGrand and R.D. Salvatore (eds), *Close Encounters of Empire: Writing the Cultural History of US-Latin America Relations* (Durham: Duke University Press, 1998), pp. 69–106.

3 See, for example J. Farley, *Bilharzia: A History of Imperial Tropical Medicine* (New York: Cambridge University Press, 1991); M. Worboys, 'Tropical Diseases', in W.F. Bynum and R. Porter (eds), *Companion Encyclopedia of the History of*

Medicine, vol. 1 (New York: Routledge, 1993), pp. 512–36; D. Arnold, 'Medicine and Colonialism', in Bynum and Porter (eds), *Companion Encyclopedia*, pp. 1393–416; A. Bashford, *Imperial Hygiene: A Critical History of Colonialism, Nationalism and Public Health* (London: Palgrave, 2004).

4 See W. Anderson, 'Immunities of Empire: Race, Disease, and the New Tropical Medicine, 1900–1920', *Bulletin of the History of Medicine*, 70 (1996): 94–118; A. Spielman and M. D'Antonio, *Mosquito: The Story of Man's Deadliest Foe* (New York: Hyperion, 2001).

5 See M.D. Gorgas and B.J. Hendrick, *William Crawford Gorgas: His Life and Work* (New York: Doubleday, 1924).

6 US Surgeon General to W.C. Gorgas, 21 December 1898, Folder '1898, November–December', Box 2, William C. Gorgas Papers, Ac. 2,409, Library of Congress, Washington, DC.

7 See J. Duffy, *The Sanitarians: A History of American Public Health* (Urbana-Champaign: University of Illinois Press, 1992).

8 M. Humphreys, *Yellow Fever and the South* (Baltimore: Johns Hopkins University Press, 1992); J.H. Ellis, *Yellow Fever and Public Health in the New South* (Lexington: University of Kentucky Press, 1992).

9 See N. Stepan, 'The Interplay between Socio-Economic Factors and Medical Science: Yellow Fever Research, Cuba and the United States', *Social Studies of Science*, 8 (1978): 397–423; for an excellent account of the Reed Commission experts see A.E. Truby, *Memoir of Walter Reed: The Yellow Fever Episode* (New York: Paul B. Hoeber, 1943).

10 See the articles: W. Reed, J. Carroll, A. Agramonte and J. Lazear, 'The Etiology of Yellow Fever – A Preliminary Note', *Proceedings of the Twenty-eighth Annual Meeting of the American Public Health Association in Indianapolis*, 22–26 October 1900; W. Reed, J. Carroll and A. Agramonte, 'The Etiology of Yellow Fever: An Additional Note', *Proceedings of the Pan-American Medical Congress in Havana*, 4–7 February 1901; W. Reed, J. Carroll and A. Agramonte, 'Experimental Yellow Fever', *American Medicine*, 6 July 1901: 15–23; W. Reed and J. Carroll, 'The Etiology of Yellow Fever: A Supplemental Note', *American Medicine*, 22 February 1902, pp. 301–5; on French colonial medicine and yellow fever research in Brazil see I. Löwy, 'Yellow Fever in Rio de Janeiro and the Pasteur Institute Mission (1901–1905): the Transfer of Science to the Periphery', *Medical History*, 34 (1990): 144–63.

11 See D.M. Haynes, *Imperial Medicine: Patrick Manson and the Conquest of Tropical Disease* (Philadelphia: University of Pennsylvania Press, 2001).

12 D. McCullough, *The Path Between the Seas: The Creation of the Panama Canal, 1870–1914* (New York: Simon & Shuster, 1977), p. 418.

13 *Ibid.*

14 Walter C. Reed to William Crawford Gorgas, June 5, 1901, Walter A.C. Reed Papers, MS C 48, National Library of Medicine, Bethesda, Maryland.

15 See N.L. Stepan, *Picturing Tropical Nature* (Ithaca: Cornell University Press, 2001).

16 W.C. Gorgas, 'Sanitation on the Canal Zone', *Journal of the American Medical Association*, 60 (29 March 1913): 954.

17 See McCullough, *The Path Between the Seas*.

18 See M.L. Conniff, *Panama and the United States: The Forced Alliance* (Athens: University of Georgia Press, 1992); J. Lindsay-Poland, *Emperors in the Jungle: The Hidden History of the US in Panama* (Durham: Duke University Press, 2003).

19 M.L. Conniff, *Black Labor on a White Canal, Panama, 1904–1981* (Pittsburgh: University of Pittsburgh Press, 1985); also see P. Sutter, '"Pulling the Teeth of the

Tropics": Environment, Disease, Race, and the US Sanitary Program in Panama, 1904–1914', unpublished manuscript in author's possession.

20 See McCullough, *Path Between the Seas* and Conniff, *Black Labor on a White Canal.*

21 Quoted in J.A. LePrince and A.J. Orenstein, *Mosquito Control in Panama: The Eradication of Malaria and Yellow Fever in Cuba and Panama* (New York: G.P. Putnam's Sons, 1916), p. 272.

22 Isthmian Canal Commission, *Laws of the Canal Zone Isthmus of Panama* (Washington, DC: Government Printing Office, 1906).

23 W.C. Gorgas, *Sanitation in Panama* (New York: D. Appleton and Company, 1915), p. 147.

24 *Ibid.*, p. 182.

25 *Ibid.*, p. 156.

26 'Report of the Department of Health of the Isthmian Canal Commission for the Month of September 1906' (Washington, DC: Government Printing Office, 1906), p. 3.

27 See 'Report of the Department of Health of the Isthmian Canal Commission for the month of January 1906' (Washington, D.: Government Printing Office, 1906).

28 W.C. Gorgas, 'Annual Report of the Department of Health of the Isthmian Canal Commission for the Year 1906' (Washington, DC: Government Printing Office, 1907).

29 'Yellow Fever: Discovery of the Mosquito Theory – Method of Controlling and Suppressing the Disease', offprint *Canal Record*, 2 February 1910, Folder 37-H-34, Part IV, Box 257, Record Group (RG 185), Isthmian Canal Commission (ICC), National Archives and Records Administration (NARA), College Park, Maryland.

30 W.C. Gorgas, 'Recent Experiences of the United States Army with Regard to Sanitation of Yellow Fever in the Tropics', *Journal of Tropical Medicine*, 6 (February 1903): 49–52.

31 H.R. Carter, 'Notes on the Sanitation of Yellow Fever and Malaria, From Isthmian Experience', *Medical Record* (19 July 1909): 2 (read at the Section of Hygiene, Pan-American Scientific Congress, Santiago, Chile, December, 1908).

32 C.F. Mason, Chief Health Officer, to Governor of the Panama Canal, 1 September 1914, Box 252, 37-F-154, RG 185, ICC, NARA.

33 See W. Anderson, '"Where Every Prospect Pleases and Only Man in Vile": Laboratory Medicine as Colonial Discourse', in V.L. Rafael (ed.), *Discrepant Histories: Translocal Essays on Filipino Cultures* (Philadelphia: Temple University Press, 1995), pp. 83–112; W. Anderson, 'Excremental Colonialism: Public Health and the Poetics of Pollution', *Critical Inquiry*, 21 (1995): 640–69; R. Ileto, 'Cholera and the Origins of the American Sanitary Order', in Rafael, *Discrepant Histories*, pp. 51–81; M. Tapper, 'Interrogating Bodies: Medico-Racial Knowledge, Politics, and the Study of a Disease', *Comparative Study of Studies in Society and History*, 37 (1995): 76–93.

34 Carter, 'Notes on the Sanitation of Yellow Fever and Malaria', p. 12.

35 R. Ross, *Mosquito Brigades and How to Organize Them* (New York City: Longmans, Green and Co., 1902), p. 50.

36 J.C. Perry, 'Colon', *Public Health Reports* (4 March 1904): 352.

37 Perry, 'A study of the vital statistics as regards prevailing diseases and mortality, of Colon, Republic of Panama, for the year 1903', *Public Health Reports* (18 March 1904): 467–75.

38 W.G. Baetz, 'Syphilis in Colored Canal Laborers – A Resume of 500 Consecutive Medical Cases', *Proceedings of the Medical Association of the Isthmian Canal Zone*, Vol. VII, Part 1 (Mount Hope, C.Z.: Panama Canal Press, 1916), pp. 17–33.
39 McCullough, *The Path Between the Seas*, p. 501; Conniff, *Black Labor on a White Canal*.
40 'Report of the Department of Health of the Isthmian Canal Commission for the month of March 1909' (Washington, DC: Government Printing Office, 1909), p. 6.
41 See Alexandra Minna Stern, *Eugenic Nation: Faults and Frontiers of Better Breeding in Modern America* (Berkeley: University of California Press, 2005), chapter 1.
42 For the best discussion of the limits of the eradication model see M. Cueto, 'Sanitation from Above: Yellow Fever and Foreign Intervention in Peru, 1919–1922', *Hispanic American Historical Review*, 72 (1992): 1–22.
43 L.W. Hackett to W. Rose, 29 August 1914, Folder 127, Box 9, Record Group 5, Series 1, Sub-series 2, Panama (327), Rockefeller Archive Center (RAC), Tarrytown, NY.
44 Rockefeller Foundation (RF), International Health Board (IHB), *Annual Report*, 1921, p. 133.
45 'A Sequel to Cuba and Panama', RF, IHB, *Annual Report*, 1919, p. 11.
46 Cueto, 'Sanitation from Above', 5.
47 J. Farley, *To Cast out Disease: A History of the International Health Division of the Rockefeller Foundation (1913–1951)* (New York: Oxford University Press, 2004), p. 92.
48 *Ibid.*, p. 88.
49 W.C. Gorgas, 'Rockefeller Foundation Health Board, Itinerary of Major General William C. Gorgas, Chairman of Yellow Fever Commission to South America, June 14–December 12, 1916', 4 July 1916, Record Group 12, Series 1, Sub-series Dairies, RF, RAC.
50 'Fighting Mosquitoes with Fish', RF, IHB, *Annual Report*, 1921, pp. 200–5.
51 Gorgas, *Sanitation in the Tropics*, 288–9.
52 D.B. Cooper and K.F. Kiple, 'Yellow Fever', in Kiple (ed.), *The Cambridge World History of Disease* (Cambridge: Cambridge University Press, 1993), pp. 1100–7.
53 Cueto, 'Cycles of Eradication'.

4
WHO-led or WHO-managed? Re-assessing the Smallpox Eradication Program in India, 1960–1980

Sanjoy Bhattacharya

The eradication of smallpox in India would not have been possible but for the contributions of many actors. The World Health Organization headquarters (WHO HQ) in Geneva, Switzerland, and its South East Asia Regional Office, based in New Delhi, India, played an extremely prominent role. So did the health ministries of the Indian central government and state governments. All these agencies set up a series of special 'eradication units', which deployed several energetic medical and public health personnel all over the sub-continent. The Soviet Union, the United States of America, Sweden and a host of other Asian and European countries provided generous doses of aid, often on a bilateral basis, in the form of field operatives, vaccine, operating kits and money. Indian and international charitable institutions made significant contributions at crucial junctures as well.[1]

The involvement of such a great variety of workers is unsurprising considering how complicated the organization of the final stages of the Indian smallpox eradication campaign turned out to be. The country was huge, with stretches of extremely difficult terrain, often with no access to transportation links. The topography was varied and specific campaign methods had to be organized for each territorial context. Linguistic and cultural diversities were as varied. More than 20 major languages and several local dialects were spoken, and a wide variety of religious traditions and class configurations were visible in the localities of each Indian state. The administrative challenges did not end there. Many sections of the Indian population were not only often uncooperative, but also frequently openly hostile to the quest for smallpox eradication.

Even though commentaries about smallpox eradication in India frequently disagree about the value of the contributions of particular players, there is a uniformly celebrated element, which is particularly noticeable in publicity documents, official histories and memoirs. These generally also present a simplified picture of a unified campaign workforce, supposedly confident about its goals and consistently effective in the field due to its educational and technical expertise. A prime example of this is provided in the foreword written by Donald Henderson, the inspirational chief of the special Smallpox Eradication Unit set up within the WHO HQ in Geneva. In the Organization's official history of the eradication program, he declares that:

> One of the most gratifying features of this programme is the unified and effective way in which the Government of India and the World Health Organization have collaborated. At every level national and WHO staff worked shoulder to shoulder, pursuing their goal with technical competence, dedication and enthusiasm.[2]

Unsurprisingly, unpublished WHO and Government of India correspondence reveals a far more complex picture. Agencies sponsored by the WHO and the Indian government were often at loggerheads on matters of strategy in particular situations, which shows that neither administrative organization was monolithic in nature. Moreover, many officials, of different nationalities and ranks, remained skeptical about the possibility of expunging variola from the sub-continent. This even included some of the campaign's staunchest supporters within the WHO, who often privately queried many of the successes claimed in relation to the dramatic reduction of the incidence of the disease in the early 1970s.

Despite their ability to provide a more nuanced understanding of the final chapters of one of the most important international health programs in the twentieth century, these administrative complexities are often ignored.[3] This chapter shows how a multi-faceted campaign unfolded at the different levels of Indian administration; a variety of political, economic and social challenges had to be overcome through careful negotiations that generally involved an array of actors. Smallpox eradication policies were, after all, not simply designed at Geneva and New Delhi and then imposed across the sub-continent by an army of willing field workers. Instead, a careful analysis of unpublished materials reveals that strategies suggested by the WHO headquarters and the Indian central government were frequently readapted before deployment in the field, not least as officials were forced to reorient their tactics in order to make them locally acceptable. These papers also point to the important fact that the eradication drive was based on the efforts of an extremely varied workforce; while the initiatives of international workers employed by the WHO was important, the endeavors of a large number of Indian men and women, with links with local communities and a range of qualifications, proved crucial to the ultimate certification of the eradication of variola in India in 1977.

A troubled advance: The creation of the infrastructure for eradication

The WHO Health Assembly's repeated calls in the 1950s for smallpox eradication caused a lot of international attention to be focused on India, as it was a major reservoir of the disease (see Table 4.1). This critical gaze made several senior members of India's central government, including Jawaharlal Nehru, the Prime Minister, extremely uneasy. The widespread incidence of variola was considered by this modernizing crusader, with a keen sense of how India was perceived on the international stage, as a stigma; a sign that his regime's agenda of rapid reform and development was unfolding far less effectively than planned.[4] Considering the fact

Table 4.1 Smallpox cases in India and the world, 1950–1977

Year	*India*	*World*	*India/World Percentage*
1950	157,487	332,224	47.4
1951	253,332	485,942	52.1
1952	74,836	155,609	48.1
1953	37,311	90,768	41.1
1954	46,619	97,731	47.7
1955	41,887	87,743	47.7
1956	45,109	92,164	48.9
1957	78,666	156,404	50.3
1958	168,216	278,922	60.3
1959	47,109	94,603	50.4
1960	31,091	65,737	47.3
1961	45,380	88,730	51.3
1962	55,595	98,700	56.3
1963	83,423	133,003	62.7
1964	41,160	75,910	54.2
1965	33,402	112,703	29.8
1966	32,616	92,620	35.2
1967	83,943	131,418	63.9
1968	30,925	80,213	37.8
1969	19,139	52,204	35.3
1970	12,341	33,663	36.7
1971	16,166	52,794	30.6
1972	20,407	65,153	31.3
1973	88,109	135,851	64.9
1974	188,003	218,364	86.1
1975	1,436	19,278	7.5
1976	Zero	953	–
1977	Zero	3,234	–

Source: R.N. Basu, Z. Jezek and N.A. Ward, *The Eradication of Smallpox from India* (WHO/SEARO: New Delhi, 1979), 36.

that the Prime Minister was the chief executive authority in the country (the Indian President, who was elected by two houses of parliament, was merely a figurehead), Nehru's views ensured action from within the confines of the federal Ministry of Health. This took the shape of the appointment of a so-called 'Central Expert Committee' in May 1958, which was asked to put forward proposals about the best means of eradicating variola.

In many ways, however, this was a rather limited exercise – setting up a committee was one thing, getting policies implemented with bureaucratic and civilian support was quite another issue. Prime Ministerial authority at this time was relatively slight, where state-level politicians and bureaucrats were allowed significant levels of autonomy. Moreover, there was political hostility to the proposal for an organized smallpox eradication program. Even Nehru, widely regarded as a charming negotiator with a knack for rallying wide-ranging political support on the domestic front, found it extremely difficult to get rid of opposition from senior bureaucrats within the federal and state-level health ministries, who were reported as being deeply divided about the issue.[5]

Senior WHO officials in Geneva were well aware of these administrative divisions, at each level of Indian government. Despite this, they developed plans for eradication for India, on the assumption that bureaucratic and political opposition in the country would ultimately be overcome with the support of senior members of the central government. This assessment was powerfully underlined by the presentation of a 'Smallpox Eradication Criterion' in August 1961.[6] Strikingly, this prescription proved unpopular within the headquarters of the WHO's South East Asia Regional Office, based in New Delhi, the Indian capital. They demanded changes in the statement released from Geneva, which were considered necessary for reasons presented as locally pertinent.[7] Although this reminder of organizational disunity irritated the WHO HQ, its officials were nevertheless forced to provide a written reassurance to New Delhi that local epidemiological and infrastructural factors would be considered during the planning and running of an Indian eradication program.[8]

The wisdom of developing regionally relevant policy was underlined very quickly. Even as pilot smallpox eradication schemes were started in one district of each of the 22 Indian states during the course of 1960, the damaging effects of local infrastructural constraints and bureaucratic disinterest became starkly obvious. The deployment of all the pilot projects, which were intended to introduce 100 per cent vaccinal coverage, was delayed everywhere, causing much publicized WHO and central targets to be missed. Worryingly for the federal Health Ministry, these setbacks appeared as other disease eradication and control programs began to hit stormy waters, and just as their managers started demanding greater chunks of central government allocations (the flagging National Malaria Eradication Program was a good case in point, as was the troubled drive for TB control).[9]

These difficulties ensured that the structures supporting the National Smallpox Eradication Program developed far more slowly than many WHO officials had hoped. The initial burst of growth was limited to the development of a new central nodal organization based in New Delhi and this was accompanied by a round of reform of local administrative rules seeking to make state-level public health officials more answerable to their superiors in New Delhi.[10] Despite this, smallpox eradication work in the states was dogged by delays and this situation was justified by persistent references to financial difficulties.[11]

While central government financial assistance allowed the completion of most of the state-level pilot schemes, several senior central government observers were disheartened by the administrative difficulties that had emerged in almost every context. Indeed, unpublished correspondence from the second half of the 1960s shows that many powerful administrators considered this proof of the impossibility of expunging variola and began to develop plans for cutting back the National Smallpox Eradication Program budget. News of these developments set alarm bells ringing throughout the WHO, causing Donald Henderson to personally approach the Director General of Indian Health Services in February 1967. His aim, which seems to have had widespread support in Geneva, was to ensure that the Indian government continued to back the eradication goal, albeit on a new basis. Henderson suggested that all aspects of the sub-continental campaign be thoroughly reformed. This was, interestingly, not only to involve governmental structures, but also to include the relevant departments of the WHO's South East Asia Regional Office; senior WHO representatives seemed to consider it politic to accept part of the blame for the problems that were continuing to hound smallpox immunization work in India.[12]

Henderson's intervention seems to have been timely, even though he appears to have been uncertain initially about the effectiveness of his efforts, and of the public declarations of support for the Indian government made by the WHO HQ.[13] One of his letters to the American Embassy in New Delhi reported, for instance, that the Indian administrators were giving mixed messages and that Geneva had no clear idea whether the sub-continental campaign would survive the year.[14] He need not have worried. The promises of additional aid caused the Indian federal authorities to reconsider their plans of scaling back their anti-smallpox measures and led to what was widely regarded as a helpful re-shuffle of bureaucrats within the central Health Ministry department charged with the responsibility for co-ordinating the eradication program.[15]

Expansion, re-organization and re-deployment: Indian Smallpox Eradication Programs, 1967–1980

The developments of 1967 brought about a major shift in the organization of the sub-continental smallpox eradication program. It is important to remember, however, that this was not just a result of the WHO decision to embark on a worldwide campaign of mass immunization. The changes initiated in Geneva and New Delhi from the month of April onwards were a direct response to the threat of withdrawal of the Indian government's support, and these brought forth heightened levels of WHO assistance. The Organization's willingness to commit extra resources did, of course, have certain advantages for the managers of its smallpox eradication units – they were able to extract the Indian Prime Minister's permission to launch a mass immunization drive in the sub-continent, with both sides agreeing that this work would be conducted in collaboration with federal Health Ministry officials. The new plans of action were, therefore, a result of the coming together of the strategic needs of both parties.

The WHO HQ attempted to kick start the goal of achieving countrywide mass vaccination by the employment of large numbers of foreign workers; as per agreement, they were expected to work with the local bureaucrats. Geneva also arranged for the re-organization of the smallpox eradication unit attached to the South East Asia Regional Office. Once again, this was achieved by the involvement of numerous foreign workers, with experience in managing public health projects, on a variety of short-term contracts. But, a number of problems cropped up. For one, it was difficult to find sufficient numbers of foreign staff; Henderson had trouble convincing the US government and universities to provide experienced consultants at this time.[16]

Additionally, the WHO offices in Geneva and New Delhi had to get the international workers available for work in the sub-continent cleared by the Indian authorities, which was not easily achieved. Reference to friction between short-term WHO consultants, officials working on long-term contracts for the South East Asia Regional Office and Indian bureaucrats was a common refrain in reports.[17]

All these problems combined to reduce the effectiveness of the mass vaccination drives launched in 1968; the continuing shortages of efficacious freeze-dried vaccine and operating kit did not help either. As a result, despite what the Indian government described as 'gigantic and concentrated' efforts to reformulate immunization policy between 1968 and 1970, the incidence of smallpox remained high. These circumstances exacerbated tensions between the eradication units run by the WHO and the Indian Health Ministry, as officials blamed each other's tactics. Henderson, for instance, referred to the problems existing between the 'various warring factions within the Ministry and between the Ministry and the States'.[18]

At another level, though, the high incidence of variola encouraged the formation of new alliances. This involved workers and bureaucrats who supported a shift from the goal of mass vaccination, to a new strategy of 'surveillance-containment', based on the isolation of infected people and selective vaccination of smallpox-stricken communities and 'rings' of contacts (immediate contacts targeted first, after which the scope of vaccination increased to cover a broader range of potential contacts).[19] These views did not, of course, go unchallenged. Reports frequently mentioned how variola outbreaks in the districts could throw carefully-laid plans for surveillance-containment out of gear in a situation where local bureaucrats frequently reverted to the strategy of mass vaccination.[20]

In the light of such administrative challenges, the managers of the WHO's smallpox eradication units in Geneva and Delhi made concerted efforts to gain the Indian central government's help in bringing hostile members of the federal and state health ministries and local bureaucrats into line. While published reports and official accounts of the eradication campaign are mostly silent about this tactical shift, unpublished correspondence reveals its importance for a range of senior WHO officials. It was not enough to have the stated support of the central authorities; it was now recognized that it was crucial formally to involve the federal government in mobilizing local political and bureaucratic support. Needless to say, this strategy of using central government assistance to bring state employees into line was neither easy nor always successful. Such high-level political support was inconstant and needed periodic renewal. The genius of Nicole Grasset – an inspirational French official employed by the WHO South East Asian Regional Office headquarters (SEARO HQ) – and Henderson, lay in their ability to make this possible through lobbying exercises, which were at times based on the unconventional tactic of approaching Prime Minister Indira Gandhi directly, sometimes in violation of diplomatic protocols. Gandhi's support was significant, as she was actively involved in centralizing power and was in a position to force relatively compliant State Chief Ministers to support, at least publicly, specific immunization campaigns.[21]

A good instance of this was provided in 1972; a year considered crucial within the WHO and the federal government (they believed that a concerted search of certain states was necessary at this time if eradication was to be achieved in India).[22] Central government cooperation, stoked in no small degree by support from the Prime Minister's office, caused the so-called 'smallpox endemic states', like Jammu and Kashmir and Bihar, to be searched intensively.[23] As a direct result, thorough surveillance-containment efforts, more rigorous than at any time in the past, were launched. Work was often conducted on a systematic, door-to-door basis, particularly in areas where smallpox outbreaks were confirmed – the policy was effective and even the most demanding assessments accepted that by late 1972, smallpox was only endemic in the four contiguous states of Bihar, Uttar Pradesh, West Bengal and Madhya Pradesh.[24]

It has to be said, however, that this success in limiting the scope of variola in India surprised many, both within the WHO and the Indian government, and questions were raised by those hostile to the eradication program about the reliability of the data being presented. Nevertheless, this reduction in the area of smallpox endemicity was seen as a major advance, so much so that the WHO began negotiations for the launch of an even more concentrated program of action, targeted primarily at the remaining pockets of variola. Based on an offer of even greater levels of financial and infrastructural assistance, these deliberations were successful and the so-called intensified smallpox eradication program was launched in 1973. Grasset and Henderson played an important role in negotiations with international financial donors – special funds were, for instance, made available after considerable efforts on their part by the Swedish International Development Agency.[25]

Despite the deployment of unprecedented levels of financial and technical resources, difficulties began to show up almost immediately in the running of the intensified program. Reports of numerous cases of bureaucratic opposition in the states, districts and sub-divisions threatened to sour the spirit of cooperation that appears to have developed amongst at least some senior WHO and Indian government officials. Faced with recurrent smallpox outbreaks across eastern India, accusations of inefficiency, impropriety and lack of commitment began to be traded in meetings and correspondence.[26] Grasset felt, for example, that problems were being created by officials at the level of state governments. She accused their officials of playing a double game, publicly promising help to the federal Health Ministry's and the WHO's smallpox eradication departments, but remaining non-committal in private.[27] Senior WHO officials, therefore, began to push the Indian government, from 1974 onwards, to convert the intensified program into a centrally controlled campaign; one that was politically supported by the Prime Minister's office and run by the federal Health Ministry's smallpox eradication department.[28]

Yet, this aim was not easily achieved in a situation where the Prime Minister's support fluctuated over time for reasons that are impossible to identify definitively. The important point, though, is that her commitment to the eradication goal varied, which kept senior WHO and Indian government officials supportive of smallpox eradication on the defensive. Indira Gandhi would sometimes fully endorse the aims of the campaign, release statements to that effect, and allow the WHO officials to distribute copies of these during their tours in the states.[29] She would also sometimes force senior state officials – the Chief Ministers and Health Ministers – to show similar levels of support.[30]

On other occasions, though, this encouragement appeared to all but evaporate. At one point, for instance, the federal Health Minister was permitted, almost at a whim, to freeze the number of international staff the WHO could deploy. The pressures imposed on state-level workers to co-operate with WHO teams, were often taken off at such moments; the hostility of several senior ministry officials to colleagues working within the smallpox eradication department which had close links with Grasset's and Henderson's offices contributed to these trends as well. Such patterns of inaction and hostility could prove to be administratively problematic. Apart from allowing the under-reporting of variola cases, it created a situation where surveillance–containment operations were mishandled – local workers would often carry out mass vaccinations over a limited area of about five mile radius, without any attention given to people at a high risk of infection. Workers often failed to detect people who were away from home and possibly carrying smallpox between villages. And district officials seeking to justify their inability to meet vaccination targets frequently exaggerated vaccination refusal rates.[31]

Thus, the Indian government's acceptance of the proposal, around the middle of 1974, that the running of the intensified program be fully centralized was widely celebrated within the WHO offices in Geneva and New Delhi, not least because it formally offered their smallpox eradication units the option of working in an organized manner with the federal Health Ministry. The officials attached to these agencies were now going to be allowed access to a centralized fund, built up with contributions from a range of donors and held in Geneva. These developments also allowed the creation of a new, well-organized program bureaucracy, that was distinct from the workforce attached to other disease control programs run by the federal and state Health Ministries. This bureaucracy was to be varied in composition, based not only on workers from the United States, Western and Eastern Europe and Asia (the Centers of Disease Control in the USA and the Soviet Academy of Sciences contributed several consultants to the WHO), but also the employment of local bureaucrats, Indian private medical practitioners and medical students from sub-continental colleges, who were placed on a variety of short-term contracts.[32]

Increased financial and infrastructural input did not automatically translate into success. The centrally controlled intensified smallpox eradication program was not always able to attract the support of local administrative networks and operate without impediment. The special status accorded to the campaign and its workforce often made it deeply unpopular amongst sections of Indian central and state government. This even included elements within the federal Health Ministry, who continued to undermine the intensified program. A dramatic example was Dr J.B. Shrivastav, the Director General of Health Services, the senior-most bureaucrat within the federal Health Ministry. Dr Mahendra Dutta, a senior member of the ministry's smallpox eradication department, noted that Shrivastav began to question the surveillance-containment policy at a time that it was considered crucial. Presenting himself as a supporter of the policy of 100 per cent vaccinal coverage, he began distributing warnings about the dangers arising from the development of a 'vaccination backlog'. This created doubts amongst more junior state- and district-level officials, who began to worry about what would happen to their career prospects if they were found to be ignoring the views of Shrivastav and his allies. They often tended, as a result, to be less than cooperative to the smallpox eradication teams.[33]

The problems did not end there. Certain state administrators began to demand that workers attached to other vertical public health programs and the health centers return to their original duties, rather than buttress the intensified program. Indeed, WHO officials soon found themselves competing for resources with family planning schemes launched by the central and state Health Ministries due to pressures imposed by Sanjay Gandhi, the Prime Minister's politically powerful son.[34] Senior WHO officials, like Grasset and Henderson, tried to lighten the impact of these developments by directing diplomatic initiatives at the Prime Minister's Office, the State Chief Ministers, and the federal and state health ministries. However, only some of these efforts proved successful; the intensified eradication program moved ahead in fits and starts during the course of 1974.[35]

Nevertheless, efforts at strengthening the program continued apace right through 1975. This took several forms: more foreign consultants were brought in from a variety of countries, greater numbers of local workers were contracted on a temporary basis with funds held at the WHO HQ, and the support of senior politicians and bureaucrats was lobbied continuously. These efforts paid off; eastern India was systematically and intensively searched for variola pockets, leading to the discovery of several cases in January that year. While the month had started off well, with less than 100 outbreaks being reported throughout the country in the first two weeks, a search carried out by a team led by Dr R.B. Arnold, a CDC epidemiologist posted to Nalanda, Bihar, revealed a large cluster of new cases at Pawa Puri village. The situation was complicated by the fact that several hundred Jain pilgrims – a religious community averse to vaccination – were visiting the village on a daily basis.[36]

Ironically, however, this outbreak proved useful to the program managers. Reference to the crisis allowed them to reinvigorate support for eradication, as several senior politicians and bureaucrats were reminded that the battle against variola was far from won. This event also allowed Grasset and Henderson, and their allies within the federal Health Ministry, to get the Prime Minister's ear, in a situation where she did not want her regime to be identified with the failure of a global program. Her office began involving itself in bringing the Congress-run state ministries into line. The benefits of such trends were clearly visible in the weeks following the Bihar outbreak. Even though several ministerial employees and civil servants doubted that variola could be expunged in the sub-continent, Indira Gandhi's firm intervention ensured that they were forced to support efforts to contain the outbreak and carry out detailed searches of surrounding areas. In this regard, the rôle played by Sharan Singh, Bihar's Chief Secretary was very important. He pressured branches of the state administration and the Chief Minister and helped ensure the deployment of governmental resources for special epidemiological teams, which were set up in association with the smallpox eradication unit in New Delhi. Singh also negotiated a political arrangement where Dr Larry Brilliant, an American consultant employed by the WHO's South East Asia Regional Office, was allowed to take over responsibility for co-ordinating activity in Pawa Puri. The central government even cleared the Bihar Military Police to assist these special epidemiological teams; military personnel helped cordon off affected villages and provided protection to program staff.[37]

Notably, the managers of the intensified program kept reminding the central and state governments, as well as national and international funding agencies, about the possibility of another serious smallpox outbreak if their work slackened. The dangers arising from such potential crises were also underlined; India, it was frequently pointed out, could very well end up bearing the stigma of causing the failure of a high profile global eradication campaign. By all indications, these tactics were effective. Funding bodies, like the Tata Industrial Group and Swedish International Development Agency, renewed their financial commitment. Surveillance-containment measures elsewhere were retained as well, generally with active assistance from the Government of India, which allowed its anti-malaria and family planning units to be used frequently by the managers of the smallpox eradication program, most notably to strengthen search activities in eastern and north-eastern parts of the country.[38]

Announcement of the so-called 'smallpox zero status' followed soon after; the last indigenous case was reported on 17 May 1975, from the Katihar district of Bihar.[39] The news was announced officially by Dr Karan Singh, the federal Minister of Health, on 30 June 1975 and then widely publicized. The achievement was also celebrated through a variety of public functions, some coinciding with the country's independence day celebrations on 15 August 1975. Even though the managers of the intensified program participated in these celebrations, they were extremely uncertain privately about the wisdom of announcing such a 'victory'.[40] Henderson and Grasset highlighted the need to push through the message that the eradication of smallpox in India could by no means be taken as guaranteed.[41] A great deal of effort was therefore expended by the WHO and the smallpox eradication department of the federal Health Ministry advertised the importance of continuing detailed countrywide searches through 1976 and 1977. These paid off, but despite Indira Gandhi's enthusiasm for this final drive, the Indian administrative services were by no means united in their support for the retention of an intensified program. Even at this late stage, when a victory against variola in India had been confidently announced by the federal authorities, many officials in New Delhi and the states still believed that the disappearance of variola was temporary and that the disease would inevitably reappear, after being re-introduced from Bangladesh or Africa.[42]

As it transpired, these fears were misplaced. It should not be forgotten that program workers of all ranks worried incessantly about unearthing a large pocket of smallpox; this anxiety even caused generous monetary rewards to be offered for the notification of variola cases and this was followed up by the detailed investigation of all resultant reports.[43] In any case, managers of the intensified program were able to start preparing the documentation that was to certify the eradication of smallpox in India by September 1976.[44] This evidence was cross-checked by an independent team of international workers over the course of several months and India was certified smallpox free on 23 April 1977.

Conclusion

The successful outcome of the smallpox eradication program demanded persistent hard work by a range of Indian and international officials. The personal sacrifices were often great: program officials were forced to spend protracted periods of time away from their families in unfamiliar contexts, and put in demanding shifts in the field that often led to physical exhaustion and ill-health. This work was frequently a thankless task, as workers encountered the hostility not only of those they were seeking to protect from a dreadful disease, but also that of politicians and officials. As a result, their experience was often bittersweet: sometimes extremely frustrating but also greatly gratifying, especially when certification of eradication was achieved in the face of overwhelming odds.[45]

A detailed examination of the experiences of these workers and of their interactions with different governmental departments and officials thus presents a nuanced picture of the multi-faceted smallpox eradication program. This program is sometimes simplistically presented as a vertically-organized campaign that was imposed on India by powerful industrialized nations. If anything, Indian administrators accepted the launch of an organized effort aimed at expunging variola on their own terms; the campaign was also run on their terms over several years, despite the best efforts of certain WHO officials to dictate the design and unfolding of policy in the sub-continent.

These trends were visible at all levels of administration. The Prime Minister, the federal Health Ministry and central government bureaucracy reminded WHO representatives of their autonomy at every available opportunity. As a result, WHO was forced to change its strategic plans for smallpox eradication in the sub-continent and agree to contribute generously to the setting up of a special bureaucracy for the purpose. And yet, this did not solve all their problems. State- and district-level administrators, keen to demonstrate their unwillingness to be ordered about by international and New Delhi-based officers, also provided differing levels of cooperation to the plans put forward by the smallpox eradication units. As a result, senior WHO officials remained acutely aware that none of their goals would be met without political and bureaucratic assistance from the highest levels of Indian government. It was recognized that such support was most likely to arise from supplicatory requests, made through diplomatic initiatives.

It is also important to note that the smallpox eradication program had variable effects on the running of the health-delivery systems based at the different levels of Indian administration. While it is undeniable that some dispensary facilities were affected adversely by the eradication drive, as health personnel were drawn away from their daily responsibilities, this situation was by no means common. In fact, accusations that the smallpox eradication program harmed the provision of local health care facilities were frequently exaggerated and politically motivated. Apart from representing the annoyance of bureaucrats and politicians doubting the possibility of eradicating variola, these criticisms were often used to deflect from the fact that many sub-divisional health care facilities were not as comprehensive as state government officials had claimed in their reports, publicity materials and election speeches.

The smallpox eradication program thus appears to have competed far more vigorously for financial resources than other centrally administered vertical health schemes, like the family planning campaigns. It is also worth noting, in this regard, that the managers of the smallpox eradication program considered it useful to employ members of local communities on short-term contracts for special anti-epidemic measures and state-level intensive surveillance-containment campaigns. These short-term employees were seen as an invaluable source of locally pertinent information, as well as useful for introducing teams of touring officials to the rural communities being targeted. These temporary workers were also asked to report on the effective working of local medical and public health officials, who were expected to notify all rash and fever cases they encountered during the course of their routine duties, for further investigation.

It is impossible to tell the complete story of a complex public health program like the smallpox eradication campaign through published WHO and Indian government reports, and the celebratory official histories and memoirs of field workers. Such commentaries tend to present an oversimplified sense of unity of purpose, over-emphasize the contributions of certain organizations and individuals, and downplay many of the serious problems bedeviling the campaign. A careful analysis of unpublished correspondence, on the other hand, shows us how policies developed at the level of the WHO HQ, and how Indian central government had to re-adapt continuously to meet local conditions. It also reveals that a range of workers of different nationalities and with widely varying professional qualifications, were responsible for a monumental triumph that many had thought impossible.

Notes

1 The WHO's official history of the last phase of the smallpox eradication program provides a detailed list of the international and Indian epidemiologists deployed in the sub-continent. See R.N. Basu, Z. Jezek and N.A. Ward, *The Eradication of Smallpox from India* (New Delhi: WHO/SEARO, 1979).
2 *Ibid.*
3 The result is the presentation of huge generalizations about the structures and goals of the World Health Organization and the smallpox eradication bureaucracy it helped set up. See for instance, footnote 21 in H. Naraindas, 'Care, Welfare, and Treason: The advent of vaccination in the 19th century', *Contributions to Indian Sociology (n.s)*, 32 (1998): 94.
4 See for instance, speech by J. Nehru about the importance of smallpox eradication reproduced in *National smallpox eradication programme in India* (New Delhi: Ministry of Health and Family Planning, Government of India, 1966), p. 1.
5 *Report of the Smallpox Worker's Conference (18–20 February 1964)* (New Delhi: Ministry of Health, 1964), p. 25.
6 Memorandum from Dr W. Bonne, Director, Communicable Diseases Section, WHO HQ, Geneva, to Dr C. Mani, Regional Director, SEARO, New Delhi, 8 August 1961, File SPX-1, Box 545, Smallpox Eradication Archives, World Health Organization, Geneva, Switzerland [hereafter, WHO/SEP].
7 Memorandum from Regional Director, SEARO, New Delhi, to the Director, Communicable Diseases Section, WHO HQ, Geneva, 18 August 1961, File SPX-1, Box 545, WHO/SEP.
8 Memorandum from Dr W. Bonne, Director, Communicable Diseases Section, WHO HQ, Geneva, to Dr C. Mani, Regional Director, SEARO, 14 September 1961, File SPX-1, Box 545, WHO/SEP.
9 See for instance, *Report 1961–6* (Delhi: Ministry of Health, Government of India, 1964), p. 3, Shastri Bhavan Library, New Delhi [hereafter, SBL].
10 See for instance, circular letter from Deputy Secretary, Ministry of Health and Family Planning, Government of India [hereafter, GOI], to Public Health Departments of all state governments, 3 January 1962, reproduced in *National Smallpox Eradication Programme*, pp. 20–3.
11 *Report 1961–62*, p. 19.
12 Letter from D.A. Henderson, Chief, Smallpox Eradication Programme, WHO HQ, Geneva, to Dr K.M. Lal, GOI, 21 February 1967, File 416, Box 193, WHO/SEP.
13 See for instance, telegram from A.M.H. Payne, WHO HQ, Geneva, to the Regional Director, SEARO, 4 April 1967, File 416, Box 193, WHO/SEP.
14 Letter from D.A. Henderson, Chief, Smallpox Eradication, WHO HQ, Geneva, to Dr E.S. Tierkel, USAID Office, American Embassy, New Delhi, 12 April 1967, File 416, Box 193, WHO/SEP.
15 See for example, letter from C. Mani, Director, SEARO, to the Ministry of Health and Family Planning, GOI, 2 May 1967, File 416, Box 193, WHO/SEP.

16 Letter from D.A. Henderson, Chief, Smallpox Eradication, WHO HQ, to the Surgeon General, United States Public Health Service, Bethesda, 6 September 1967, File 416, Box 193, WHO/SEP.
17 Personal letter from Dr N. Maltseva, SEARO, to D.A. Henderson, Chief, Smallpox Eradication, WHO HQ, 27 June 1967, File 416, Box 193, WHO/SEP.
18 Personal letter from D.A. Henderson, Chief, Smallpox Eradication, WHO HQ, to Dr A. Oles, SEARO, 14 September 1970, File 416, Box 193, WHO/SEP.
19 For a reference to this, see personal letter from D.A. Henderson, Chief, Smallpox Eradication Unit, WHO HQ, to Dr J. Brown, Los Angeles, 12 May 1970, File 416, Box 193, WHO/SEP.
20 See for instance, personal letter from D.A. Henderson, Smallpox Eradication Unit, WHO HQ, to Dr V.A. Muhopad, Lucknow, 28 June 1971, File 416, Box 193, WHO/SEP.
21 See letter from Dr N. Grasset, Regional Advisor, Smallpox Eradication, SEARO, to D.A. Henderson, Chief, Smallpox Eradication, WHO HQ, 15 September 1972, File 830, Box 194, WHO/SEP.
22 Letter from Dr N. Grasset, Regional Advisor, Smallpox Eradication, SEARO, to Mrs I. Gandhi, PM, India, 14 September 1972, File 830, Box 194, WHO/SEP.
23 M. Singh, 'Report on a visit to study the implementation of the National Smallpox Eradication Programme in Jammu & Kashmir (12th November to 27th November 1972)', c.1973, p. 1, File 830, Box 194, WHO/SEP. See also J.M. Pifer, 'Confidential: Smallpox in Bihar State of India During 1971' (c.1971), pp. 11–12, File 436, Box 193, WHO/SEP.
24 Personal letter from D.A. Henderson, Chief, Smallpox Eradication, WHO HQ, to Dr N. Grasset, Regional Advisor, Smallpox Eradication, SEARO, 26 September 1972, File 830, Box 194, WHO/SEP.
25 See letter from N.K. Jungalwalla, GOI, to Dr D.A. Henderson, Chief, Smallpox Eradication, WHO HQ, 12 December 1973, File 948, Box 17, WHO/SEP.
26 Personal letter from D.A. Henderson, Chief, Smallpox Eradication, WHO HQ, to Dr N. Grasset, Regional Advisor, Smallpox Eradication, SEARO, 5 March 1973, File 830, Box 194, WHO/SEP.
27 Personal and confidential letter from Dr N. Grasset, Regional Advisor, Smallpox Eradication, SEARO, to D.A. Henderson, Chief, Smallpox Eradication, WHO HQ, Geneva, 7 June 1973, File 830, Box 194, WHO/SEP.
28 For examples of such official trends, see File 830, Box 194, WHO/SEP.
29 See for example, Indira Gandhi's October 1974 statement and the publications released in support by state-level officials, see File 832, Box 197, WHO/SEP.
30 *Ibid.*
31 'Review on the situation of reporting of smallpox, investigation and containment of outbreaks, Bhopal Division', c.1974, pp. 1–4, File 832, Box 197, WHO/SEP.
32 For a good description of the experiences of American workers in South Asia, see P. Greenough, 'Intimidation, Coercion and Resistance in the final stages of the South Asian Smallpox Eradication Campaign, 1973–75', *Social Science and Medicine*, 41 (1995): 633–45.
33 M. Dutta, 'Snakes and Ladders: An untold story of the fight against smallpox in India', unpublished TS, c.1980, p. 9, Private papers of Mahendra Dutta. Also, interview with Dr M. Dutta, New Delhi (Delhi), India, 3 March 1999.
34 Dutta, 'Snakes and Ladders', p. 13.
35 Letter from Dr N. Grasset, Regional Advisor, Smallpox Eradication, SEARO, to D.A. Henderson, Chief, Smallpox Eradication Unit, WHO HQ, c. September 1974, File 388, Box 194, WHO/SEP.

36 Dutta, 'Snakes and Ladders', 9–11.

37 *Ibid.*

38 See for instance, memorandum from SEARO, to D.A. Henderson, Chief, Smallpox Eradication, WHO HQ, c. August 1975, File number 831, Box 195, WHO/SEP.

39 See letter from H. Mahler, Director General, WHO, to K. Singh, Minister of Health and Family Planning, GOI, 20 August 1975, File number 831, Box 195, WHO/SEP. Also see personal letter from L.B. Brilliant, Medical Officer, SEARO, New Delhi, to WHO HQ, 20 August 1975, File number 831, Box 195, WHO/SEP.

40 See letter from V.T.H. Gunaratne, Regional Director, SEARO, to H. Mahler, Director General, WHO, 1 July 1975, File number 831, Box 195, WHO/SEP. Singh's TV broadcast was immediately – and widely – reported in the Indian press. See for example, 'Smallpox wiped out, says Karan Singh' in *Indian Express* (1 July 1975); 'Small-pox eradicated, claims minister', *Times of India* (1 July 1975); 'No smallpox in India, *Hindustan Times* (1 July 1975), and 'Smallpox eradicated', *National Herald* (1 July 1975).

41 See for instance, letter from D.A. Henderson, Chief, Smallpox Eradication, WHO HQ, to Z. Jezek, Medical Officer, SEARO, 3 November 1975, File number 831, Box 195, WHO/SEP.

42 Personal letter from N. Grasset, Regional Advisor, Smallpox Eradication, SEARO to I. Gandhi, Prime Minister, India, 16 February 1976, File 831, Box 195, WHO/SEP.

43 Letter from L.B. Brilliant, Medical Officer, SEARO, to A.K. Chakravarty, Government of West Bengal, 3 December 1975, File 832, Box 197, WHO/SEP.

44 Letter from D.A. Henderson, Chief, Smallpox Eradication, WHO HQ, to L.B. Brilliant, Michigan, USA, c. September 1976, File number 831, Box 195, WHO/SEP.

45 Letter from N. Grasset, Regional Advisor, Smallpox Eradication, SEARO, to D.A. Henderson, Chief, Smallpox Eradication, WHO HQ and V.T.H. Gunaratne, Regional Director, SEARO, 30 June 1975, File number 831, Box 195, WHO/SEP.

5

The World Health Organization and the Transition from 'International' to 'Global' Health

Theodore M. Brown, Marcos Cueto and Elizabeth Fee

Even a quick glance at the titles of books and articles in recent medical and public health literature suggests that an important transition is underway. The terms 'global', 'globalization', and their variants are everywhere, and in the specific context of international public health, 'global' seems to be emerging as the preferred authoritative term.[1] As one indicator, the number of entries under the rubrics 'global' and 'international' health in PubMed shows that 'global' health is rapidly on the rise, seemingly on track to overtake 'international' health at some point in 2005 (see Table 5.1). Although universities, government agencies, and private philanthropies are all using the term in highly visible ways, the origin and meaning of the term 'global health' are still unclear.[2]

The purpose of this chapter is to provide historical insight into the emergence of the terminology of 'global health'. We believe that an examination of this linguistic shift will yield important fruit, and not only about fashions and fads in language use. Our task is to provide a critical analysis

Table 5.1 Total number of PubMed entries under international* and global* by decade

Keyword Searches, January 2005	*International**	*Global**
1950s	1,007	54
1960s	3,303	155
1970s	8,369	1,137
1980s	16,924	7,176
1990s	49,158	27,794
2000–2004	44,372⁺	33,621⁺

* Picks up variant term endings, e.g. internationalize and internationalization; globalize and globalization
⁺ NB 48 months only

of the meaning, emergence, and significance of the term 'global health' and to place its growing popularity in a broader historical context. A particular focus of this paper is the role of the World Health Organization (WHO) in both 'international' and 'global' health and as an agent in the transition from one term to the other.

Let us first define and differentiate some essential terms. 'International' health was already a term of considerable currency in the late nineteenth and early twentieth century and referred primarily to a focus on the control of epidemics across the borders or boundaries between nations, that is 'inter-nationally'. 'Intergovernmental' refers to the relationships between the governments of sovereign nations, in this case with regard to the policies and practices of public health. 'Global' health means different things to different people but in general implies the consideration of the health needs of the people of the whole planet as an agenda above the concerns of particular nations. The term 'global' makes no distinction among nation-states and other social constituents of society and is thus associated with and recognizes the growing importance of actors beyond governmental or intergovernmental organizations and agencies – for example, the press and media, internationally influential foundations, non-governmental organizations, transnational corporations. Logically, the terms 'international', 'intergovernmental' and 'global' need not be mutually exclusive and in fact can be understood as complementary. Thus we could say that the WHO is an intergovernmental agency that plays international roles and exercises international functions – with the goal of improving global health.

Given these definitions, it should come as no surprise that 'global' health is not entirely an invention of the past few years. 'Global' was sometimes used (in the context of international health) well before the 1990s, as in the 'global malaria eradication program' launched by WHO in the mid-1950s, the title of a WHO Public Affairs Committee pamphlet of 1958, 'The World Health Organization: Its Global Battle Against Disease',[3] the 1971 report on 'The Politics of Global Health' for the US House of Representatives,[4] and many studies of the 'global population problem' in the 1970s.[5] But the term was generally limited and its use in official statements and documents was sporadic at best. Many believe that we are now living in a different era and facing a new reality for public health: hence the increasing frequency of references to 'global' health.[6] Yet the questions remain: how many have participated in this shift in terminology and do they consider it trendy, trivial, or trenchant?

These questions attracted the attention of Supinda Bunyavanich and Ruth B. Walkup who published, under the provocative title 'US Public Health Leaders Shift Toward a New Paradigm of Global Health', their report of conversations conducted in 1999 with 29 'international health leaders'.[7] Their respondents divided into two groups. About half felt there was no need for a new terminology and that the label 'global health' was meaningless jargon.

The other half thought there were profound differences between 'international' and 'global' health and that 'international' clearly meant coordination constrained by national boundaries and interests, whereas 'global' just as clearly meant something transnational. Although these respondents felt that a major shift had occurred within the previous few years, they seemed unable clearly to articulate or define it.

In 1998, Derek Yach and Douglas Bettcher came closer to capturing both the essence and the origin of the new 'global health' in a two-part article on 'The Globalization of Public Health' published in the *American Journal of Public Health*.[8] They defined the 'new paradigm' of globalization as 'the process of increasing economic, political, and social interdependence and integration as capital, goods, persons, concepts, images, ideas and values cross state boundaries'. The roots of globalization were long, they said, going back at least to the nineteenth century, but the process was assuming a new magnitude, unprecedented in world history, in the late twentieth century. The globalization of public health, they argued, had a dual aspect, one promising and one threatening.

On the up side, they argued, was easier diffusion of appropriate and useful technologies and of ideas and values – such as human rights. On the down side, were such risks as diminished social safety nets, the facilitated marketing of tobacco, alcohol, and psychoactive drugs, the easier worldwide spread of infectious diseases, global warming, and the rapid degradation of the environment with dangerous public health consequences. But Yach and Bettcher were convinced that WHO could turn these risks into opportunities. Countries could collaborate in 'global intersectoral action through transnational cooperation and partnerships', and adopt 'more comprehensive forms of global vigilance, research and monitoring'. The WHO could help create more efficient information and surveillance systems by 'strengthening its global monitoring and alert systems' thus creating 'global early warning systems'. Even the most powerful nations would buy into this new globally interdependent world system once they realized that such involvement was in their best interest. In the face of global disease threats, enlightened self-interest and altruism would converge.

Despite the long list of problems and threats, Yach and Bettcher were largely uncritical as they promoted the virtues of global public health and the central leadership role of WHO. In an editorial in the same issue of the Journal, George Silver, professor of public health at Yale University, noted that Yach and Bettcher worked for WHO and that their stance was similar to other optimistic and self-promotional stances taken by WHO officials and advocates. But WHO, Silver pointed out, was actually in a bad way. 'The WHO's leadership role', he wrote, 'has passed to the far wealthier and more influential World Bank, and the WHO's mission has been dispersed among other UN agencies'.[9] UN agencies did what they could with restricted resources because wealthy donor countries were billions of dollars in arrears, and this left the UN and its agencies in 'disarray, hamstrung by financial constraints and internal incompetencies, frustrated by turf wars and cross-national policies'. Given these realities, Yach and Bettcher's promotional stance on 'global public health', while affiliated with the WHO, was, to say the least, intriguing. Why were these spokesmen for the much-criticized and apparently hobbled WHO so upbeat about 'global' public health?

WHO: the early years

To better understand Yach and Bettcher's role, and that of the WHO more generally, it will be helpful briefly to review the history of the organization from 1948 to 1998, as it moved from being the unquestioned leader of international health to facing the challenge of finding its place in the contested world of 'global health'.

The WHO formally began in 1948 when the first World Health Assembly in Geneva ratified its constitution. The idea of a permanent institution for international health can be traced to the organization in 1902 of the International Sanitary Office of the American Republics which, some decades later, became the Pan American Sanitary Bureau and eventually the Pan American Health Organization.[10] Also important in the early twentieth century was the Rockefeller Foundation and its International Health Division, which tackled the problems of hookworm, yellow fever, and malaria and exported throughout the world a U.S.-inspired model of medical education and public health practice.[11]

Two European-based international health agencies were also important. One was the Office International d'hygiène publique, which functioned in Paris from 1907, concentrating on several basic activities: the study of epidemic diseases; preparing international sanitary conferences; revising and administering international sanitary agreements; and the rapid exchange of epidemiological information.[12] The second agency, the League of Nations Health Organization, began its work in 1920, as Zylberman discusses in Chapter 2.[13] This organization established its headquarters in Geneva, Switzerland, and sponsored a series of international conferences and commissions, published epidemiological intelligence and technical reports, and promoted the study of malaria, cancer, leprosy, and such new fields as rural and industrial hygiene. Between 1921 and 1939 the Medical Director of the Hygiene Section, Ludwik Rajchman, emphasized social medicine, and pointed to worldwide social pathologies such as poverty, inadequate housing, and malnutrition as the root causes everywhere of illness and premature death.[14] The League of Nations Health Organization was poorly budgeted and faced covert opposition from other national and international organizations, including the US Public Health Service. Despite these complications, which limited the League's effectiveness, both the

Office international d'hygiène publique and the Health Organization survived through World War II, and were present at the critical post-war moment when the future of international health would be defined.

An international conference in San Francisco in 1945 approved the creation of the United Nations and also voted for the creation of a new specialized health agency. Participants at the San Francisco meeting initially formed a commission of 16 prominent individuals from around the world, among whom were René Sand from Belgium, Andrija Stampar from Yugoslavia, and Thomas Parran from the United States. Sand and Stampar were widely recognized as champions of social medicine. The commission held meetings between 1946 and early 1948 to establish the new international health organization. Representatives of the Pan American Sanitary Bureau, whose leaders resisted being absorbed by the new agency, were also involved, as were leaders of new institutions such as the United Nations Relief and Rehabilitation Administration (UNRRA) that had been working since 1943 on urgent humanitarian issues in Europe.

Against this background, the first World Health Assembly met in Geneva in June 1948 and formally created the World Health Organization. The Paris Office, the League of Nations Health Organization, and UNRRA merged into the new agency. The Pan American Sanitary Bureau – then headed by Fred L. Soper, a former Rockefeller Foundation official – was allowed to retain autonomous status as part of a regionalization scheme.[15] WHO formally divided the world into a series of regions: the Americas, Southeast Asia, Europe, Eastern Mediterranean, Western Pacific, and Africa, but did not fully implement this regionalization until the 1950s. An 'international' and 'intergovernmental' mindset generally prevailed in the 1940s and 1950s but calling the new organization the World Health Organization also raised sights to a worldwide, 'global' perspective. This was underscored in the Universal Declaration of Human Rights, adopted and proclaimed on December 10, 1948 by the United Nations General Assembly. Article 25 of the Declaration begins as follows: 'Everyone has the right to a standard of living adequate for the health and well-being of himself and his family, including food, shelter, housing, and medical care and necessary social services...'

The first Director General of WHO, Brock Chisholm, was a Canadian psychiatrist loosely identified with the British social medicine tradition. The United States, a main contributor to the WHO budget, played a contradictory role: on the one hand, it supported the United Nations system with its broad worldwide goals, but on the other, it was jealous of its own sovereignty and maintained the right to be the sole arbiter of 'inter-American' affairs and to intervene unilaterally in the Americas in the name of 'national security'. Another problem for WHO was that its constitution had to be ratified by nation-states, a slow process: by 1949 only 14 countries had signed on.[16]

As an intergovernmental agency, WHO had to be responsive to the larger political environment. The politics of the Cold War had a particular salience, with an unmistakable impact on WHO policies and personnel. Thus when the Soviet Union and other communist countries, including mainland China, walked out of the UN system and therefore out of WHO in 1949, the United States and its allies were easily able to exert a dominating influence. In 1953, Chisholm completed his term and was replaced by the Brazilian Marcolino Candau. Candau had worked under Soper on malaria control in Brazil and was associated with the 'vertical' disease control programs of the Rockefeller Foundation, and then with their adoption by the Pan American Sanitary Bureau when Soper moved to that agency as Director.[17] Candau would be Director General of WHO for over 20 years, being reelected to four successive 5-year terms. He was a practical man and did what had to be done. During the period between 1949 and 1956, at which time the Soviet Union returned to the UN and WHO, WHO was closely allied with the United States, US interests, and US foreign policy.

In 1955, Candau was charged with overseeing WHO's campaign of malaria eradication, approved by the Western-weighted World Health Assembly in 1955. As Randall Packard has argued, the United States and its allies understood that global malaria eradication was a question of economic and political development as much as public health.[18] Malaria eradication was intended to usher in economic growth and create overseas markets for US technology and manufactured goods. It would build support for local governments and their US supporters, and help win 'hearts and minds' in the battle against Communism. The campaign reproduced the development theories of the time by promoting technologies brought in from outside, valuing the knowledge of foreign experts, and making little or no attempt to enlist the participation or cooperation of local populations in planning or implementation. This model of development assistance fitted neatly into US Cold War efforts to promote modernization with limited social reform.[19]

Over time, WHO shifted its position on malaria eradication. With the return of the Soviet Union and other communist countries in 1956, the political balance in the World Health Assembly also shifted and Candau, as WHO's Director General, had to accommodate the changed balance of power. At the same time, in the field, malaria eradication had at first seemed to work but it was facing serious difficulties; ultimately it would suffer colossal and embarrassing failures. In 1969 the World Health Assembly declared that it was not feasible to eradicate malaria in many parts of the world and began a slow process of reversal, returning once again to an older, less ambitious, malaria control agenda. Scientific and political factors both played a role in that reversal. This time, however, there was a new twist; the 1969 Assembly emphasized the need to develop rural health systems and to integrate malaria control into general health services.

When the Soviet Union returned to WHO in 1956, its representative at the Assembly was Viktor Zhdanov, Deputy Minister of Health of the USSR. He argued that it was now scientifically feasible, socially desirable, and economically worthwhile to attempt to eradicate smallpox worldwide.[20] The USSR perhaps wanted to make its mark on global health and Candau, recognizing the shifting balance of power, was willing to collaborate. The USSR agreed to provide 25 million doses of freeze-dried vaccine and Cuba, 2 million doses of vaccine, and in 1959, the World Health Assembly committed itself to a global smallpox eradication program.

In the 1960s, technical improvements – jet injectors and bifurcated needles – made the process of vaccination much cheaper, easier, and more effective. The United States interest in smallpox eradication sharply increased, and in 1965, Lyndon Johnson instructed the US delegation to the World Health Assembly to pledge American support for an international program to eradicate smallpox from the earth.[21] Despite a decade of progress, the disease was still endemic in more than 30 countries, with 10 to 15 million annual cases, and some 2 million deaths. In 1967, now with the support of the world's most powerful players, WHO launched the Intensified Smallpox Eradication Program. This intensified eradication program, an international effort led by the American Donald A. Henderson, would ultimately be stunningly successful.[22]

The promise and perils of primary health care, 1973–1993

Within WHO, there have always been tensions between the social and economic analysis of ill-health – and therefore the social and economic solutions to population health problems – and technologically or disease-focused approaches to health. These tendencies are not necessarily incompatible, although they are often at odds. Certainly the emphasis on one or the other waxes and wanes over time, depending on the larger balance of power, the specific changing interests of different international players, the intellectual and ideological commitments of key individuals, and the way that all of these factors impact on the health policy-making process.

During the 1960s and 1970s, changes in WHO were driven by a political context marked by the emergence of decolonized African nations, the spread of nationalist and socialist movements, and new proposals for development that emphasized long-term socioeconomic development instead of short-term technological interventions. Rallying in organizations such as the Non-Aligned Movement, developing countries created the United Nations Conference on Trade and Development (UNCTAD) where they argued vigorously for fairer terms of trade and the more liberal financing of development.[23] In Washington, DC, more liberal political shifts succeeded the conservative tenor of the 1950s, with the civil rights movement and other social movements forcing changes in national priorities.

This changing political environment was reflected by corresponding shifts within WHO. In the 1960s, WHO acknowledged that a health infrastructure was a prerequisite for the success of the malaria control program, especially in Africa. In 1968, Candau called for a comprehensive health plan, within which an integrated approach to curative and preventive care could be developed. A Soviet representative to the executive board called for an organizational study of methods for promoting the development of basic health services.[24] In January 1971, the executive board agreed to undertake the organizational study and this was presented to the full executive board in 1973.[25] Socrates Litsios has discussed many of the steps in the transformation of WHO's approach from an older model of health services research to what would become the 'Primary Health Care' approach.[26] This drew upon the thinking and experiences of the Christian Medical Commission of the World Council of Churches and the experiences of non-governmental organizations (NGOs) and medical missionaries working in Africa, Asia, and Latin America at the grass-roots level. It also gained saliency from China's reentry into the UN in 1973 and the widespread interest in, and popular enthusiasm about, the Chinese 'barefoot doctors' who were reported to be transforming health conditions in rural areas. These experiences underscored the urgency of a 'Primary Health Care' perspective that included the training of community health workers and the solution of basic economic and environmental problems.[27] To some extent, these proposals returned to the social medicine priorities and 'world' perspective priorities that had animated the original WHO.

These new tendencies were now embodied in WHO by Halfdan T. Mahler, a Dane, who served as Director General from 1973 to 1988. Under considerable pressure from the Soviet delegate to the executive board, Dimitri Venediktov, Mahler finally agreed to hold a major conference on the organization of health services in Alma-Ata, in the Soviet Union. Mahler was initially reluctant because he disagreed with the Soviet's highly centralized, over-medicalized approach to the provision of health services.[28] And although the Soviet Union succeeded in having the conference on its territory, the resulting conference, held in September 1978, reflected Mahler's view much more closely than it did that of the Soviets. The Declaration of Primary Health Care and the goal of 'Health for All in the year 2000' advocated an 'inter-sectoral' and multidimensional approach to health and development, emphasized the use of 'appropriate technology', and urged active community participation in health care and health education at every level. The Declaration of Alma-Ata concluded: 'A genuine policy of independence, peace, détente and disarmament could and should release additional resources that could well be devoted to... the acceleration of social and economic development of which primary health care, as an essential part, should be allotted its proper share'.[29]

David Tejada has argued that 'It is regrettable that afterward the impatience of some international agencies, both UN and private, and their emphasis on achieving tangible results instead of promoting change – something that is always difficult – led to major distortions of the original concept of primary health care'.[30] For many governments, agencies, and individuals, WHO's idealistic view of Primary Health Care was indeed perceived as 'unrealistic'. The process of reducing WHO's idealistic view to a practical set of technical interventions that could more easily be implemented, managed, and measured, began in 1979 at a small conference, with a heavy US flavor, held in Bellagio, Italy, and sponsored by the Rockefeller Foundation, with assistance from the World Bank. Those in attendance included the President of the World Bank, the Vice President of the Ford Foundation, the Administrator of USAID, and the Executive Secretary of UNICEF.[31] The Bellagio meeting focused on an alternative concept to that articulated at Alma-Ata – 'Selective Primary Health Care' – which was built on the notion of pragmatic, low cost technical interventions that were limited in scope and easy to monitor and evaluate. Thanks primarily to UNICEF, Selective Primary Health Care was soon operationalized under the acronym 'GOBI' (Growth monitoring to fight malnutrition in children, Oral rehydration techniques to defeat diarrheal diseases, Breastfeeding to protect children, and Immunizations).[32]

Before this conference, the WHO had already initiated the Expanded Program on Immunization in 1974.[33] This program selected six diseases – tuberculosis, diphtheria, neonatal tetanus, whooping cough, poliomyelitis, and measles – which each had an effective and inexpensive vaccine, and mounted major immunization campaigns around the world. Until this time, only about 5 per cent of children in many developing countries received immunizations.[34] But now, the success of smallpox eradication gave the Expanded Program on Immunization new glamor and legitimacy, and provided a substantial boost to the Selective Primary Health Care agenda.

In the 1980s, WHO had to reckon with the growing influence of the World Bank. The Bank had initially been formed in 1946 to assist in the reconstruction of Europe after World War II, and later expanded its mandate to provide loans, grants, and technical assistance to developing countries. At first, it funded large investments in physical capital and infrastructure, but then, in the 1970s, began to invest in population control, health, and education, with the emphasis on population control.[35] The World Bank approved its first loan for family planning in 1970 and its first loan for nutrition in 1976. In 1979, the Bank created a Population, Health, and Nutrition Department and adopted a policy of funding both standalone health programs and health components of other projects.

In its 1980 *World Development Report*, the Bank argued that both malnutrition and ill-health could be addressed by direct government action – with Bank assistance.[36] It also suggested that improving health and nutrition could accelerate economic growth, thus providing a good argument for social sector spending. Throughout the 1980s the Bank awarded loans and grants for food and nutrition, family planning, maternal and child health, and basic health services. As the Bank began to make direct loans for health services, it called for more efficient use of available resources and discussed the roles of the private and public sectors in financing health care. The Bank favored free markets for financing health care, and a diminished role for national governments.[37] In the context of widespread developing country indebtedness and increasingly scarce resources for health expenditures, the World Bank's promotion of structural adjustment measures and cuts in social and health expenditures – at the very time that the HIV/AIDS epidemic erupted – drew angry criticism of the Bank but also underscored its new influence.

In contrast to the World Bank's growing influence in international health, in the 1980s the prestige of the World Health Organization was beginning to diminish. Its policies and operations were still caught in the cross-fire of Cold War politics. One sign of trouble was the 1982 vote by the World Health Assembly to freeze WHO's budget.[38] This was followed by the 1985 decision by the United States to pay only 20 per cent of its assessed contribution to all UN agencies and to withhold its contribution to WHO's regular budget, in part as a protest against WHO's 'Essential Drug Program' which was opposed by leading US-based pharmaceutical companies.[39] These events occurred amidst growing tensions between WHO and UNICEF and other agencies and the increasingly divisive controversy over Selective versus Comprehensive Primary Health Care. As part of a rancorous public debate conducted in the pages of *Social Science and Medicine* in 1988, Kenneth Newell, a highly placed WHO official and an architect of Comprehensive Primary Health Care, called Selective Primary Health Care a 'threat ... [that] can be thought of as a counter-revolution'.[40]

In 1988, Mahler's 15-year tenure as Director General of WHO came to an end. Unexpectedly, Hiroshi Nakajima, a Japanese researcher with experience in drug evaluations, who had been Director of the WHO Western Pacific Regional Office in Manila, was elected as the new Director General.[41]

Crisis at WHO, 1988–1998

The first citizen of Japan ever elected as head of a UN agency, Nakajima rapidly became the most controversial Director General in WHO's history. His nomination had not been supported by the United States or by a number of European and Latin American countries, and his performance in office did little to assuage their doubts. Nakajima did try to launch several important initiatives – on tobacco, global disease surveillance, and public-private partnerships – but fierce criticisms persisted and raised questions about his autocratic style and poor management, his inability to communicate effectively, and worst of all, cronyism and corruption. Among his other problems, Nakajima had an open dispute with Jonathan Mann, a respected American physician who had organized WHO's Global Program on AIDS and who was recognized as a pioneer in raising world AIDS awareness.[42] Mann resigned suddenly and dramatically.

Another symptom of WHO's problems in the late 1980s was the growth of extra-budgetary funding. As Gill Walt of the London School of Hygiene and Tropical Medicine noted, there was a crucial shift from predominant reliance on WHO's 'regular budget' – drawn from member states' contributions based on population size and GNP – to greatly increased dependence on 'extra-budgetary' funding coming from donations by multilateral agencies or 'donor' nations.[43] In 1971 the regular budget was $75 million and the extra-budgetary portion was $25 million. By 1986–1987, extra-budgetary funds of $437 million had almost caught up with the regular budget of $543 million. By the beginning of the 1990s, extra-budgetary funding had overtaken the regular budget by $21 million, thus contributing 54 per cent of WHO's overall budget. Enormous problems for the Organization followed from this budgetary shift. Priorities and policies were still ostensibly set by the World Health Assembly, which was made up of all member nations, increased from 55 in 1948 to 178 in 1992. But the World Health Assembly, now dominated numerically by poor and developing countries, had authority only over the regular budget, frozen, in any case, since the early 1980s. Wealthy donor nations and multilateral agencies like the World Bank could largely call the shots on the use of the extra-budgetary funds they contributed. They thus created, in effect, a series of 'vertical' programs more or less independent of the rest of WHO's programs and decision-making structure. The WHO was loath to turn these funds away. The dilemma for the Organization was that although the extra-budgetary funds added to the overall budget, 'they increase difficulties of coordination and continuity, cause unpredictability in finance, and a great deal of dependence on the satisfaction of particular donors'.[44]

Fiona Godlee, assistant editor at the *British Medical Journal*, published an aggressive series of articles in 1994–1995 that built on Walt's critique. These carried such titles as 'WHO in crisis', 'WHO in retreat: is it losing its influence', and 'WHO's special programmes: undermining from above'.[45] She concluded with this dire assessment: 'WHO is caught in a cycle of decline, with donors expressing their lack of faith in its central management by placing funds outside the management's control. This has prevented WHO from coordinating its activities in line with centrally agreed priorities and has undermined attempts to develop integrated responses to countries' long term needs'.

In the late 1980s and early 1990s, the World Bank moved confidently into the vacuum created by an increasingly ineffective WHO. WHO officials were unable or unwilling to respond to the new international political economy structured around neo-liberal approaches to economics, trade, and politics.[46] The Bank was well adapted to this role, and was relatively insulated from the concerns of populations in developing countries. The Bank maintained that existing health systems were often wasteful, inefficient, and ineffective, and argued in favor of greater reliance on private-sector health care provision, and the reduction of public involvement in health services delivery.[47]

Controversies surrounded the Bank's policies and practices, but there was no doubt that, by the early 1990s, it had become a dominant force in international health. The Bank's greatest 'comparative advantage' lay in its ability to mobilize large financial resources; by 1990, the Bank's loans for health surpassed the total budget of WHO, and by the end of 1996, the Bank's cumulative lending portfolio in health, nutrition, and population had reached $13.5 billion. The Bank's other advantage was its ability to exert pressure upon relatively powerful Ministries of Finance and Planning. WHO, on the other hand, dealt mainly with Ministries of Health – among the least powerful government sectors. Yet the Bank recognized that, whereas it had great economic strengths and influence, WHO still had considerable technical expertise in matters of health and medicine. This was clearly reflected in the Bank's widely influential World Development Report of 1993, *Investing in Health* in which credit is given to WHO, 'a full partner with the World Bank at every stage in the preparation of the Report'.[48] Circumstances suggested that it was to the advantage of both parties for the Bank and WHO to work together.

WHO embraces 'global health'

This is the context in which the WHO began to refashion itself as a coordinator, strategic planner, and leader of 'global health' initiatives. In January 1992, the 31-member Executive Board of the World Health Assembly decided to appoint a 'working group' to recommend how WHO could be most effective in international health work in the light of the 'global change' rapidly overtaking the world. The Executive Board may have been responding, in part, to the Children's Vaccine Initiative, perceived within the WHO as an attempted 'coup' by UNICEF, the World Bank, the UN Development Program, the Rockefeller Foundation and several other players seeking to wrest control of vaccine development.[49] The working group's final report of May 1993 recommended that WHO – if it were to maintain leadership of the health sector – must strengthen its capacity in technical analysis, overhaul the fragmented management of global, regional, and country programs, diminish the competition between regular and extra-budgetary programs, and above all, increase the emphasis within WHO on global health issues and WHO's coordinating role in that domain.[50]

Until that time, the term 'global health' had been used sporadically and, outside the WHO, usually by people on the political left with various 'world' agendas. G.A. Gellert of International Physicians for the Prevention of Nuclear War published an article in 1990 calling for analyses of 'global health interdependence'.[51] In the same year, Milton and Ruth Roemer published a paper on 'Global health, national development, and the role of government', arguing that further improvements in global health would be dependent on the expansion of public rather than private health services.[52] Another strong source for the term 'global health' was the environmental movement and especially debates over world environmental degradation, global warming, and their potentially devastating effects on human health.[53]

In the mid-1990s a considerable body of literature developed on global health threats and emerging diseases. In the United States, a new CDC journal, *Emerging Infectious Diseases*, began publication and former CDC director William Foege started using the phrases 'global infectious disease threats' and 'global microbial threats'.[54] In 1997 the Institute of Medicine's Board of International Health released a report revealingly entitled *America's Vital Interest in Global Health: Protecting Our People, Enhancing Our Economy, and Advancing Our International Interests*.[55] In 1998 the CDC's *Preventing Emerging Infectious Diseases: A Strategy for the 21st Century* appeared, followed in 2001 by the Institute of Medicine's *Perspectives on the Department of Defense Global Emerging Infections Surveillance and Response System*.[56] Bestselling books and news magazines were full of stories about Ebola and West Nile virus, resurgent tuberculosis, and the threat of bioterrorism.[57] The message was clear: there was a palpable global disease threat.

In 1998, the World Health Assembly reached outside the ranks of WHO for a new leader who could restore credibility to the organization and provide it with a new vision: Gro Harlem Brundtland, a former Prime Minister of Norway, and a physician and public health professional. Brundtland brought formidable expertise to the task. In the 1980s she had been the Chairperson of the United Nations World Commission on Environment and Development and produced the 'Brundtland Report' which led to the Earth Summit of 1992. She was familiar with the debates over global warming and climate change and the global thinking of the environmental movement. She had a broad and clear understanding of the links between health, environment, and development, and had connections with the major players in these fields.[58]

Brundtland decided on a major reorganization of WHO's management structure. She created a Cabinet of nine members, five of them women, and only three selected from within WHO. Following the principle of 'One WHO' she made efforts to streamline the bureaucratic structure, abolish redundant posts, reduce barriers between regions, bring in new managers from developing countries, and create a more transparent system of accountability and disclosure.[59] Many but not all of her efforts at internal reform succeeded.

Brundtland was determined to position WHO as an important player on the global stage, to move beyond ministries of health and gain a seat at the table when decisions were being made.[60] She wanted to refashion WHO as a 'department of consequence' able to monitor and influence the health impact of other actors on the global scene. She established a Commission on Macroeconomics and Health, chaired by the economist Jeffrey Sachs of Harvard University, and including former ministers of finance, and officers from the World Bank, the International Monetary Fund, the World Trade Organization, and the United Nations Development Program, as well as public health leaders. The Commission issued a report in December 2001, which argued that improving health in developing countries was essential to their economic development.[61] The report identified a set of disease priorities that would require focused intervention. Notably, it reaffirmed the role of the state in providing services in poor and middle-income countries, and cautioned against reckless privatization.

Brundtland also began to strengthen WHO's financial position, largely by organizing 'global partnerships' and 'global funds' to bring together 'stakeholders' – private donors, governments, and bilateral and multilateral agencies – to concentrate on specific targets (for example, Roll Back Malaria in 1998, GAVI in 1999, and Stop TB in 2001). These were semi-autonomous programs bringing in substantial outside funding, often in the form of Public-Private Partnerships.[62] A very significant player was the Bill & Melinda Gates Foundation, which committed more than $1.7 billion between 1998 and 2000 to an international program to prevent or eliminate diseases in the world's poorest nations, mainly through vaccines and immunization programs.[63] Within a few years, some 70 'global health partnerships' had been created.

Brundtland's tenure as Director General was not without blemish nor free from criticism. It has been noted that some of the initiatives credited to her administration had actually been started under Nakajima (for example, the WHO Framework Convention on Tobacco Control), that others may be looked upon today with some skepticism (the Commission on Macroeconomics and Health, Roll Back Malaria), and that still others did not receive enough attention from her administration (primary health care, HIV/AIDS, health and human rights, and child health). Nonetheless, few would dispute the assertion that Brundtland succeeded in achieving her principal objective, which was to reposition WHO as a credible and highly visible contributor to the rapidly changing field of global health.

Conclusion

We can now return briefly to the questions with which this chapter began: how does a historical perspective help us understand the emergence of the terminology of 'global health' and what role did the WHO play as an agent in its development? The basic answer is that WHO at various times in its history alternatively led, reflected, and tried to accommodate to broader changes and challenges in the ever-shifting world of international health. In pursuing a path of institutional survival in the 1990s, WHO used leadership of an emerging concern with 'global health' as a useful organizational strategy that promised survival, and indeed, renewal. The WHO did not invent 'global health'; other, larger forces were responsible for that. But the WHO certainly helped promote interest in the field and contributed significantly to the dissemination of new concepts and a new vocabulary. Whether the WHO gained substantially in the process is still an open question. Whether WHO's process of institutional repositioning will contribute to the health of the world's population – the deeper question – also remains unanswered at this time.

Notes

The views expressed in this paper are those of the authors and not necessarily those of the institutions with which they are associated.

1 A small sampling of recent titles: D.L. Heymann and G.R. Rodier, 'Global surveillance of communicable diseases', *Emerging Infectious Diseases*, 4 (1998): 362–5; D. Woodward, N. Drager, R. Beaglehole and D. Lipson, 'Globalization and Health: A Framework for Analysis and Action', *Bulletin of the World Health Organization*, 79 (2001): 875–81; G. Walt, 'Globalisation of International Health', *Lancet*, 351 (7 February 1998): 434–7; S.J. Kunitz, 'Globalization, States, and the Health of Indigenous Peoples', *American Journal of Public Health*, 90 (2000): 1531–9; K. Lee, K. Buse and S. Fustukian (eds), *Health Policy in a Globalising World* (Cambridge: Cambridge University Press, 2002).

2 For example, Yale has a Division of Global Health in its School of Public Health, Harvard has a Center for Health and the Global Environment, and the London School of Hygiene and Tropical Medicine has a Center on Global Change and Health; the National Institutes of Health has a strategic plan on Emerging Infectious Diseases and Global Health; Gro Harlem Brundtland addressed the 35th Anniversary Symposium of the John E. Fogarty International Center on 'Global Health: A Challenge to Scientists' in May 2003; the Centers of Disease Control and Prevention has established an Office of Global Health and has partnered with the World Health Organization, the World Bank, UNICEF, the United States Agency for International Development, and others, in creating Global health Partnerships.

3 A. Deutsch, *The World Health Organization: Its Global Battle Against Disease* (New York: Public Affairs Committee, 1958).

4 R.M. Packard, '"No Other Logical Choice": Global Malaria Eradication and the Politics of International Health in the Post-War Era', *Parassitologia*, 40 (1998): 217–29 and *The Politics of Global Health*, prepared for the Subcommittee on National Security Policy and Scientific Developments of the Committee on Foreign Affairs, US House of Representatives (Washington, DC: US Government Printing Office, 1971).

5 For example, T.W. Wilson, *World Population and a Global Emergency* (Washington, DC: Aspen Institute for Humanistic Studies, Program in Environment and Quality of Life, 1974).

6 J.E. Banta, 'From International to Global Health', *Journal of Community Health*, 26 (2001): 73–6.

7 S. Bunyavanich and R.B. Walkup, 'US Public Health Leaders Shift Toward a New Paradigm of Global Health', *American Journal of Public Health*, 91 (2001): 1556–8.

8 D. Yach and D. Bettcher, 'The Globalization of Public Health, I: Threats and Opportunities', *American Journal of Public Health*, 88 (1998): 735–8; and Yach and Bettcher, 'The Globalization of Public Health, II: The Convergence of Self-Interest and Altruism', *American Journal of Public Health*, 88 (1998): 738–41. The exceptions were Milton and R. Roemer, 'Global Health, National Development, and the Role of Government', *American Journal of Public Health*, 80 (1990): 1188–92.

9 G. Silver, 'International Health Services Need an Interorganizational Policy', *American Journal of Public Health*, 88 (1998): 728.

10 Organización Panamericana de la Salud. *Pro Salute, Novi Mundi: Historia de la Organización Panamericana de la Salud* (Washington, DC: OPS, 1992).

11 See J. Farley, *To Cast Out Disease: A History of the International Health Division of the Rockefeller Foundation (1913–1951)* (Oxford: Oxford University Press, 2003); A-E. Birn, 'Eradication, Control or Neither? Hookworm Versus Malaria Strategies and Rockefeller Public Health in Mexico', *Parassitologia*, 40 (1996): 137–47; M. Cueto (ed.) *Missionaries of Science: Latin America and the Rockefeller Foundation* (Bloomington: Indiana University Press, 1994).

12 Office international d'hygiène publique, *Vingt-cinq ans d'activité de l'Office international d'hygiène publique*, 1909–1933 (Paris: Office international d'hygiène publique, 1933); P.F. Basch, 'A Historical Perspective on International Health', *Infectious Disease Clinics of North America*, 5 (1991): 183–96; and W.R. Aykroyd, 'International Health – A Retrospective Memoir', *Perspectives in Biology and Medicine* (1968): 273–85.

13 F.G. Bourdreau, 'International Health', *American Journal of Public Health and the Nation's Health*, 19 (1929): 863–78; Bourdreau, 'International Health Work' in H.E. Favis (ed.), *Pioneers in World Order: An American Appraisal of the League of Nations* (New York: Columbia University Press, 1944), pp. 193–207; N. Howard-Jones, *International Public Health Between the Two World Wars: The Organizational Problems* (Geneva: WHO, 1978); and M.D. Dubin, 'The League of Nations Health Organisation' in P. Weindling (ed.) *International Health Organisations and Movements, 1918–1939* (Cambridge: Cambridge University Press, 1995), pp. 56–80.

14 M. Balinska, 'For the Good of Humanity: Ludwik Rajchman', *Medical Statesman* (New York: Central European University Press, 1998); P. Weindling, 'Social Medicine at the League of Nations Health Organisation and the International Labour Office Compared' in Weindling (ed.), *International Health Organisations and Movements*, pp. 134–53.

15 T. Parran, 'The First 12 Years of WHO', *Public Health Reports* 73 (1958): 879–83; F.L. Soper, *Ventures in World Health: The Memoirs of Fred Lowe Soper*, J. Duffy (ed.), (Washington DC: PAHO, 1977); and J. Siddiqi, *World Health and World Politics: The World Health Organization and the UN System* (London: Hurst and Co., 1995).

16 'Seventh Meeting of the Executive Committee of the Pan American Sanitary Organization', Washington, DC, 23–30 May 1949. Folder 'Pan American Sanitary Bureau', R.G. 90–41, Box 9. Series Graduate School of Public Health, University of Pittsburgh Archives.

17 World Health Organization, 'Information. Former Directors-General of the World Health Organization. Dr. Marcolino Gomes Candau' http://www.who.int/archives/who50/en/directors.htm (accessed 24 July 2004); 'In Memory of Dr. M.G. Candau', *WHO Chronicle*, 37 (1983): 144–7.

18 R.M. Packard, 'Malaria Dreams: Postwar Visions of Health and Development in the Third World', *Medical Anthropology*, 17 (1997): 279–96; R. Packard, 'No Other Logical Choice'.
19 R.M. Packard and P.J. Brown, 'Rethinking Health, Development and Malaria: Historicizing a Cultural Model in International Health', *Medical Anthropology*, 17 (1997): 181–94.
20 I. Glynn and J. Glynn, *The Life and Death of Smallpox* (New York: Cambridge University Press, 2004), pp. 194–6.
21 *Ibid.*, p. 198.
22 W.H. Foege, 'Commentary: Smallpox Eradication in West and Central Africa Revisited', *Bulletin of the World Health Organization*, 76 (1998): 233–5; D.A. Henderson, 'Eradication: Lessons From the Past', *Bulletin of the World Health Organization*, 76 (1998): Suppl 2: 17–21; F. Fenner, D.A. Henderson, I. Arita, Z. Ježek and I. Dalinovich Ladnyi, *Smallpox and Its Eradication* (Geneva: WHO, 1988).
23 J.N. Bhagwati (ed.), *The New International Economic Order: The North South Debate* (Cambridge, Mass: MIT Press, 1977); R.L. Rothstein, *Global Bargaining: UNCTAD and the Quest for a New International Economic Order* (Princeton, N.J.: Princeton University Press, 1979).
24 S. Litsios, 'The Long and Difficult Road to Alma-Ata: A Personal Reflection', *International Journal of Health Services*, 32 (2002): 709–32.
25 WHO, Executive Board 49th Session, document EB49/SR/14 Rev. (Geneva: WHO, 1973), p. 218; WHO, 'Organizational study of the Executive Board on methods of promoting the development of basic health services' (Geneva: WHO, 1972), document EB49/WP/6, 19–20.
26 S. Litsios, 'The Christian Medical Commission and the Development of WHO's Primary Health Care Approach', *American Journal of Public Health*, 94 (2004): 1884–93; S. Litsios, 'The Long and Difficult Road to Alma-Ata'.
27 J.H. Bryant, *Health and the Developing World* (Ithaca: Cornell University Press, 1969); C.E. Taylor (ed.), *Doctors for the Villages: Study of Rural Internships in Seven Indian Medical Colleges* (New York: Asia Publishing House, 1976); and K.W. Newell, *Health By the People* (Geneva: WHO, 1975). See also M. Cueto, 'The Origins of Primary Health Care and Selective Primary Health Care', *American Journal of Public Health*, 94 (2004): 1864–74; and S. Litsios, 'The Christian Medical Commission and the Development of WHO's Primary Health Care Approach', *American Journal of Public Health*, 94 (2004): 1884–93.
28 See Litsios, 'The Long and Difficult Road to Alma-Ata', pp. 716–19.
29 Declaration of Alma-Ata. International Conference on Primary Health Care, Alma-Ata, USSR, 6–12 September 1978. http://www.who.int/hpr/NPH/docs/declaration_almaata.pdf (accessed 10 April 2004).
30 D.A. Tejada de Rivero, 'Alma-Ata Revisited', *Perspectives in Health Magazine: The Magazine of the Pan American Health Organization*, 8 (2003): 1–6.
31 M. Black, *Children First: The Story of UNICEF, Past and Present* (Oxford: Oxford University Press, 1996) and M. Black, *The Children and the Nations: The Story of UNICEF* (New York: UNICEF, 1986), pp. 114–40. UNICEF was created in 1946 to assist needy children in Europe's war ravaged areas. After the emergency ended, it broadened its mission and concentrated resources on the needs of children in developing countries.
32 UNICEF, *The State of the World's Children: 1982/1983* (New York: Oxford University Press, 1983).

33 W. Muraskin, *The Politics of International Health: The Children's Vaccine Initiative and the Struggle to Develop Vaccines for the Third World* (Albany, NY: State University of New York Press, 1998).
34 World Health Organization, 'Immunization, Vaccines and Biologicals. The History of Vaccination' http://www.who.int/vaccines-diseases/history/history.shtml (accessed 11 April 2004).
35 J.P. Ruger, 'Changing Role of the World Bank in Global Health in Historical Perspective', *American Journal of Public Health*, 95 (January 2005): 60–70.
36 World Bank, *World Development Report 1980* (Washington, DC: World Bank, 1980).
37 World Bank, *Financing Health Services in Developing Countries: An Agenda for Reform* (Washington, DC: World Bank, 1987).
38 F. Godlee, 'WHO In Retreat; Is it Losing its Influence?' *British Medical Journal*, 309 (1994): 1493.
39 *Ibid.*, p. 1492.
40 K. Newell, 'Selective Primary Health Care: The Counter Revolution', *Social Science and Medicine*, 26 (1988): 906.
41 P. Lewis, 'Divided World Health Organization Braces For Leadership Change', *New York Times* (1 May 1988): 20.
42 P.J. Hilts, 'Leader in U.N'.s Battle on AIDS Resigns in Dispute Over Strategy', *The New York Times* (17 March 1990): 1.
43 G. Walt, 'WHO Under Stress: Implications for Health Policy', *Health Policy*, 24 (1993): 125–44.
44 *Ibid.*, p. 129.
45 F. Godlee, 'WHO In Crisis', *British Medical Journal*, 309 (1994): 1424–8; 'WHO In Retreat: Is it Losing its Influence?' *British Medical Journal*, 309 (1994): 1491–5; 'WHO's Special Programmes: Undermining From Above, *British Medical Journal*, 310 (1995): 178–82.
46 P. Brown, 'The WHO Strikes Mid-life Crisis', *New Scientist*, 153 (1997): 12; Editorial, 'World Bank's Cure For Donor Fatigue', *Lancet*, 342 (10 July 1993): 63–4; A. Zwi, 'Introduction to Policy Forum: The World Bank and International Health', *Social Science and Medicine*, 50 (2000): 167.
47 World Bank, *Financing Health Services in Developing Countries: An Agenda for Reform* (Washington, DC: World Bank, 1987).
48 World Development Report, *1993: Investing in Health* (Washington, DC: The World Bank, 1993), pp. iii–iv.
49 For a full account, see Muraskin, *The Politics of International Health.*
50 B. Stenson and G. Sterky, 'What Future WHO?' *Health Policy*, 28 (1994): 242.
51 G.A. Gellert, 'Global Health Interdependence and the International Physicians' Movement', *Journal of the American Medical Association*, 264 (1990): 610–13.
52 M. Roemer and R. Roemer, 'Global Health, National Development, and the Role of Government', *American Journal of Public Health*, 80 (1990): 1188–92.
53 See for example, A.J. Haines, 'Global Warming and Health', *British Medical Journal*, 302 (1991): 669–70; A.J. Haines, P.R. Epstein, A.J. McMichael, 'Global Health Watch: Monitoring impacts of environmental change', *Lancet*, 342 (11 December 1993): 1464–9; A.J. McMichael, 'Global Environmental Change and Human Population Health: A conceptual and scientific challenge for epidemiology', *International Journal of Epidemiology*, 22 (1993): 1–8; J.M. Last, 'Global Change: Ozone depletion, greenhouse warming, and public health', *Annual Review of Public Health*, 14 (1993): 115–36; A.J. McMichael, *Planetary Overload, Global Environmental Change and the Health of the Human Species*

(Cambridge: Cambridge University Press, 1993); A.J. McMichael, A.J. Haines, R. Sloof and S. Kovats, *Climate Change and Human Health* (Geneva: WHO, 1996); A.J. McMichael and A. Haines, 'Global Climate Change: The Potential Effects on Health', *British Medical Journal*, 315 (1997): 805–9.

54 S.S. Morse, 'Factors in the Emergence of Infectious Diseases', *Emerging Infectious Diseases*, 1 (1995): 7–15.

55 Institute of Medicine, *America's Vital Interest in Global Health: Protecting our people, enhancing our economy, and advancing our international interests* (Washington, DC: National Academy Press, 1997).

56 R.M. Krause (ed.), *Emerging Infections: Biomedical research reports* (San Diego: Academic Press, 1998); Centers for Disease Control and Prevention, Preventing Emerging Infectious Diseases: A strategy for the 21st century (Atlanta: CDC, 1998); P.S. Brachman et al. (eds), *Perspectives on the Department of Defense Global Emerging Infections Surveillance and Response System* (Washington, DC: National Academy Press, 2001).

57 For example, L. Garrett, *The Coming Plague: Newly emerging diseases in a world out of balance* (New York: Farrar, Straus and Giroux, 1994).

58 L.K. Altman, 'US Moves To Replace Japanese Head of WHO', *New York Times*, (20 December 1992): 1

59 A. Dove, 'Brundtland Takes Charge and Restructures the WHO', *Nature Medicine*, 4 (1998): 992.

60 I. Kickbusch, 'The Development of International Health Policies – Accountability Intact?' *Social Science and Medicine*, 51 (2000): 979–89.

61 Commission on Macroeconomics and Health, *Macroeconomics and Health: Investing in Health for Economic Development* (Geneva: WHO, 2001); see also H. Waitzkin, 'Report of the WHO Commission on Macroeconomics and Health: A summary and critique', *Lancet*, 361 (8 February 2003): 523–6.

62 M.A. Reid and E.J. Pearce, 'Whither the World Health Organization?' *The Medical Journal of Australia*, 178 (2003): 9–12.

63 M. McCarthy, 'A Conversation With the Leaders of the Gates Foundation's Global Health Program: Gordon Perkin and William Foege', *Lancet*, 356 (8 July 2000): 153–5.

Part II

National Security: Migration, Territory and Border Regulation

6

Where is the Border?: Screening for Tuberculosis in the United Kingdom and Australia, 1950–2000

Ian Convery, John Welshman and Alison Bashford

The modern history of the regulation of entry of non-citizens, and its connection to the emergence of nations in the nineteenth and twentieth centuries, is receiving increasing historical attention. This is particularly true of histories of nationalism and public health in North America and Australia. It has been argued that the immigration history of Australia in the twentieth century, for example, has strong links with public health policy and practice. Alison Bashford has been concerned to integrate the history of health and infectious disease control into the study of immigration and citizenship, seeing it as an element of the legal and technical constitution of 'undesirable' entrants. Part of the effect of joint infectious disease and immigration regulation over the twentieth century, she argues, has been the imagining, as well as the technical implementation, of the island-nation as ostensibly secure.[1]

The theme of the 'immigrant menace' has been similarly influential in histories of migration and public health in the US. In his important early history, *Silent Travelers*, Alan Kraut focused on the institutional response of national, state, and local governments to immigration and public health; and migrant responses to differences between their own conceptions of therapy, health care, disease prevention, and hygiene, and those of native-born Americans. In terms of disease causality, argues Kraut, stigmatism and racism ultimately obscured as much as they revealed.[2] Nick King has observed that the resurgence of tuberculosis in the US has led to renewed concern over the borders that separate people, and has created a dilemma of how to address inequalities in health while maintaining non-discriminatory policies. Essentialist narratives have assumed intrinsic differences between people, and identified diseases as coming from outside; complex problems have been reduced to single causes. Despite the emergence of more complicated social understandings of the disease in the 1990s, which stress the role of poverty, homelessness, and structural factors in the US itself, King argues that immigration has continued to be seen as the cause of the rising incidence of tuberculosis.[3] Amy Fairchild, on the other hand, has suggested that medical examinations at Ellis Island before 1924 served a more complex social function than simple exclusion on the basis of disease, class, or race, and acted as an important inclusionary mechanism, to absorb migrants into the laboring body. She has written that 'the immigrant medical examination was shaped by an industrial imperative to discipline the laboring force in accordance with industrial expectations'.[4]

In the case of the United Kingdom, evidence underlines the need to differentiate between policy responses, especially on the geographic issue of precisely where screening has taken place: this raises the spatial question (which is also a political and cultural one) about where the medico-legal border of the nation lies. One of the striking points in the debate in the UK on the tuberculosis screening of migrants in the post-World War II period was the way the emphasis on screening at the Port of Entry (as in Ellis Island for example) was subverted in favor of follow-up in the district of intended residence, what we call the 'local destination' system.[5] In the face of the arguments about importation, as well as established models (such as those in the US and Australia for screening at the border), UK government opted for a policy that was still essentially concerned with case finding, but which moved its site of operation from the Ports of Entry to local authorities that received large numbers of migrants. It is recognizably the system that has existed to the present day in the UK. In terms of national frontiers, then, the UK evidence suggests that borders could be moved internally – to the local level – as well as externally.[6]

This chapter contributes to these debates by comparing tuberculosis screening in Australia and the UK focusing on the period since 1950. It outlines differences in the two screening systems, and tries to account for diverging policy response outcomes through exploring the disparate (though connected) migration histories of the two countries. In turn, this opens up important supplementary questions. Where was the border actually located in these two geopolitical histories: offshore, at point of departure, at point of arrival, or internally at point of local destination? What political and symbolic work did the differing placement of these borders perform? Who were the 'others' in the two national contexts? And how was otherness defined and circulated, through legal, medical and media representations of migrants, asylum seekers and infectious disease? What emerges from the comparative historical material is a picture of the exceptionalism of UK policy and practice. While Australia (along with the US, Canada, New Zealand) have long histories of health-based exclusions, the UK is only now moving towards this model of public health and infectious disease regulation, with a clear emphasis on the national border. We argue in this chapter that the 'offshore' (that is pre-entry) model of screening prospective migrants has been highly symbolic, if not medically efficacious in Australian history, and is likely to function similarly in the present UK situation, in which border security and border crossings have become highly politicized.

Current health screening in Australia

What, then, is the Australian model of migrant health screening? In stark contrast to the UK, the health screening of (non-citizen) entrants to Australia is regulated less by public health law than by migration law, and has been since the nineteenth century. Currently, health screening is managed through a complex visa system, in place since the late 1980s: all people who are not citizens must gain a visa for entry (at times there have been over 200 types of visa). Except for a very few categories (for example diplomatic visas and certain emergency humanitarian visas) different health criteria are attached to the issuing of each class of visa. The nature of the health criteria, declarations, and diagnostic tests to be undertaken vary according to category of visa sought, but also according to length of stay in Australia, and risk-status of the country from which the visa is applied. Some entrants, from some countries, may simply declare their health and infectious disease status – usually on forms distributed en route. Applicants from countries deemed to be 'very high risk' who intend to stay over 3 months must pass the health criteria. And any applicant from any country who wishes to stay over 12 months – intending migrants, asylum claimants, international students, and those seeking long-term residency – must likewise undertake the health screening process: it is compulsory.[7]

For example, a UK passport-holder entering Australia (from somewhere in the British Isles) on a short-term tourist visa, must declare their health status, with particular reference to tuberculosis. For a passport-holder applying, for example, from China for a 24-month student visa, a medical examination for prescribed diseases and symptoms, and a chest X-ray, are compulsory. The X-ray must be gained in China itself, not on or after entry to Australia.[8] The visa cannot be issued unless this is done, and no one can enter the country (legally) without a visa, except, arguably, those seeking asylum. In other words, the compulsory health criteria is one of the main reasons for the visa system itself.[9]

For the purposes of illuminating both current and past UK practice, there are four points to emphasize about the processes and the substance of Australian health screening. First, this screening is compulsory under the Commonwealth Migration Act (1958). Interestingly in Australia, this compulsion currently engenders very little public or political debate, although it has in the past. Second, law and regulation recognize a 'public charge' argument, as well as a public health argument, in refusing entry to people with certain diseases and conditions. Third, and importantly in this context, what sharply distinguishes the Australian model from the UK model is that this screening and examination takes place 'offshore', in the Australian parlance. All intending migrants, international students, and long-term residents must apply for a visa, and undertake all tests and any resulting treatment in their original country-of-application.

Fourth, and related to these, tuberculosis control holds a longstanding exceptional place in Australian migration regulation: while the exclusion of a person with any other communicable disease may be wavered by Ministerial discretion, there is a 'no exception, no exemption' policy with respect to tuberculosis. An applicant with active disease must undertake chemotherapy and be re-tested before a visa will be granted.[10] The broad risk categorization of countries (where Australia is 'low risk') is determined by the incidence of tuberculosis. 'Very high risk' countries are currently (2005) listed as Algeria, Argentina, Bangladesh, Brazil, Chile, China, India, Indonesia, Korea, Malaysia, Pakistan, Papua New Guinea, Philippines, Portugal, Russia, Serbia and Montenegro, Singapore, Sri Lanka, South Africa, Vietnam and Zimbabwe. People applying from these places, whether or not they have active, inactive, or no tuberculosis, will have many more bureaucratic (and costly) hoops through which to jump, to gain an entry visa. Where a visa is applied from can make considerable difference to the required criteria and outcome. In practice, then, although not by design, this directly affects the national (and therefore the ethnic) composition of entrants to Australia. As Mawani argues of the Canadian experience, apparently neutral, race-free, policy can still be organized on race- or nationality-based principles, and can have discriminating outcomes in practice.[11] Thus whilst it is enticing to think of the present as somehow post-colonial, this 'post-coloniality' is still deeply entwined with colonial formations – a past reworked and inventively adapted in the present.[12]

Current health screening in the United Kingdom

The British Port Health Screening system is very different, because only a proportion of new entrants are accessed and referred for screening: offshore border control was historically considered, but was passed over for local follow-up. Screening operates only at international ports with Port Health Units attached, and relies upon Immigration Officers identifying certain categories of travelers and referring them to Port Health Units for screening by Port Health Officers. Screening consists of a brief medical examination and a chest X-ray. Consultants in Communicable Disease Control (CsCDC) are then notified of medical examination and chest X-ray status via standard referral forms (Port 101, 102, and 103). Port 101 covers those where medical examination and chest X-ray have been completed and the individual 'appears to be in satisfactory health'. Port 102 is issued for those for whom a chest X-ray has not been carried out. Port 103 is issued for those considered to have disease which 'may endanger the public health'. The

CsCDC are expected to arrange appropriate follow-up, usually by passing the information on to local National Health Service (NHS) tuberculosis services.[13]

The British Thoracic Society's Code of Practice recommends that all migrants (and other entrants planning to stay longer than 6 months such as university students) from all countries other than the European Union, Canada, the United States, Australia, and New Zealand, should be screened. In addition, all refugees should be screened. It suggests that an incidence of 40 cases per 100,000 population per year is an arbitrary but reasonable level above which tuberculosis might be considered common.[14] These guidelines have been endorsed by the Department of Health, and there is extensive screening at certain Ports and for certain categories of migrants. At Heathrow Airport, for example, in the period 1995–1999, 224,305 people were referred to the Health Control Unit; 169,029 migrants, and 55,276 political asylum seekers. Of the latter group, 41,470 received a chest X-ray.[15]

However in the UK there is an important disjunction between what is recommended in guidelines produced by professional bodies such as the British Thoracic Society, and what actually happens in practice. Indeed this focus on the alleged importation of disease is reflected in policy documents and certain organizational arrangements, but has never (yet) translated into the kind of offshore-screening model so established in Australian history. Despite the British Thoracic Society guidelines on the control and prevention of tuberculosis, studies have highlighted widespread variations in policies for screening of migrants. In fact, there is neither national policy nor guidelines on the extent of screening.

At the local level, many Trusts do not arrange screening for the new entrants referred to them, claiming that they do not have the resources or have other priorities. Many migrants do not attend for screening due to changes of address, language difficulties, and mistrust of authorities. In the 1980s, it was argued that the local destination (Port of Arrival) system performed poorly, with a low proportion of migrants being X-rayed or having forms completed, and a low pick up rate for clinical tuberculosis. One study estimated that in the period 1983–1988 only 36,000 of the 140,000 migrants arriving in the UK (26 per cent) were traced and screened.[16] Similarly, a study in Blackburn, for the period 1990–1994, found that only 898 of 2,242 new migrants (40 per cent) were identified through the local destination system, and 1,344 (60 per cent) through local links with the Family Health Services Authority. From this cohort, ten cases of clinical tuberculosis were found (equivalent to 450 cases per 100,000). The study concluded that the local destination system performed even more poorly in 1990–1994 than in 1983–1988, and with a low pick-up of clinical cases.[17]

A study published in 2005 set out to determine practice among CsCDC in response to Port 101 and Port 102 notification forms (new entrants who had either been screened and found to have a normal X-ray, not had a chest X-ray due to pregnancy or young age, or whose examination was inconclusive). This found that the response of CsCDC in response to the Port 101 and 102 forms varied considerably across the country. Limited capacity of services, the perception of limited benefits accruing from follow-up, combined with the absence of national guidance, meant that CsCDC took a variety of different pragmatic approaches. The system of responding to new entrants at risk of tuberculosis seemed to be incoherent. Moreover staff at the Ports appeared to interpret policy in different ways, referring differing proportions of those potentially eligible for screening.[18] Another study compared port health and tuberculosis databases in Essex, and suggested that follow-up of those with Port 101 and 102 forms should be abandoned. Only one individual was identified with tuberculosis as a direct result of port health screening.[19]

Some experts argue for a policy move away from the privileging of border screening. Richard Coker, for example, has drawn on this evidence to argue more generally that compulsory screening of migrants for tuberculosis and HIV is not based on adequate evidence, and has practical and ethical problems.[20] Nevertheless, infectious disease and migration have become increasingly important political issues. The Conservative Party argues that infection should not be imported into the UK.[21] It announced in February 2005 that under their policy, visas would be denied to prospective migrants who tested positive for tuberculosis. Michael Howard (the then party leader) stated that:

> The British people deserve the best standards of public health. We need to control who is coming to Britain to ensure that they are not a public health risk and to protect access to the NHS. It's plain common sense. And it's exactly what they do in New Zealand, Canada and Australia.[22]

According to *The Economist*, the Conservatives planned to 'import much of the Australian immigration system'.[23] And indeed, Government proposals announced in February 2005 were to implement existing powers by screening visa applicants for tuberculosis on 'high risk' routes, and requiring those diagnosed to seek treatment before they would be allowed entry to the UK.[24] Thus policy in the UK has recently shifted towards pre-entry screening for tuberculosis. The law cannot be used to remove people once they have arrived, and policies have been drawn up to exclude before arrival. This is in fact the longstanding Australian model.

History, migration, and public health

How and why did these (until recently) very different models arise historically? The answer lies in these nations' specific geopolitical histories, their histories of migration and of racialized nation-formation.

The history of migration in the modern period is also, in part, the history of colonialism: the movement and settlement of European people in other continents. Australia is the exemplary case: apart from the Indigenous community, it is a nation of immigrants, and from the beginning of the British penal colony at Sydney it was a colony of 'settlement'. In contrast to the UK, therefore, there is a strong Australian tradition of both migration law and linked quarantine law, policy and regulation stretching back into the early nineteenth century. From about the 1880s, Migration Acts in the then six British colonies (Australia was federated in 1901) became concerned with two phenomena, increasingly discursively linked: disease and race. All the Australasian colonies had some version of a Chinese Exclusion Act, part of the rationale for which was public health and communicable disease control.[25] The Immigration Restriction Act, passed by the new federal government in 1901 – the legal basis of the 'White Australia' policy – had a public health power: the 'loathsome diseases' section (3d). Thus in Australia, the connections between public health, migration, and race were explicit, legal, and technical, rationalized by the epidemiology and science of the times.[26]

In fact these kinds of racialized exclusion Acts were common to many colonies and nations of white settlement in the period, and have been linked to the nineteenth-century Chinese diaspora, and in particular to a re-emerging concern over leprosy.[27] Bashford has argued that geopolitics is part of the explanation for the particularly strong version of this international phenomenon in Australia. First, the island-status of the continent is crucial. A British (that is white) settlement in the Asia-Pacific region felt the need to assert its racial difference stridently, and constantly felt that its borders were under threat.[28] The 'flipside' of this position is the continued existence of a Fourth World (Aboriginal Australia) within the First World borders of Australia. Second, the island-status of Australia also gave rise to rigid maritime quarantine practice (still in place with respect to animals and goods in airports and seaports) in order to keep diseases out of a continent where numerous diseases were not endemic. In many respects this was successful: cholera, for example, never entered the island-continent. Historically, quarantine law provided for the detention and inspection of people. Thus on the question of compulsory detention, the connection between migration law, public health law, and policy on defense and security is very tight indeed in the Australian past and present, as Bashford and Strange have argued of current policy on detention of asylum seekers.[29]

But there is another dimension in this history, which concerns the UK specifically. Partly because of the history of race-based exclusions in Australia, the majority of migrants to the country have always been, and indeed remain, (white) British people. Through any number of formal schemes and agreements, including the Empire Settlement Act of 1922, fit and healthy British migrants were sought by the Australian government to people a nation-continent understood to be problematically underpopulated and vulnerable.[30] After World War II a number of Free and Assisted Passage schemes funded jointly by the UK and Australian governments attempted to stimulate the flow of migrants to Australia. The 'Nest Egg' scheme, for example, relaxed financial entry requirements, fares for teenagers were abolished, and assisted air passages were provided. The Australian Government also opened additional migration offices in major UK provincial cities. The net impact of these efforts was a total emigration from the UK to Australia, for the period 1946–1960, of 566,429 persons; 380,426 of whom received free or assisted passages. By 1972, when the Empire and Commonwealth Settlement Acts finally lapsed, assisted immigrants to Australia from Britain totalled 1,011,985.[31]

The original rationale of Australian migration legislation was to exclude Chinese and other 'colored aliens'. Yet their very exclusion meant that the practice of health screening 'offshore' – that is, before entry to Australia – developed over the twentieth century with respect to British applicants, British migrants. The longstanding governmental and bureaucratic links between Australia and Britain – the Colonial Office, Australia House, colonial state and federal agents-general and so on – offered the infrastructure for this process, and it built up gradually over time. For example, in the late nineteenth century, many British migrants required a vaccination certificate, or at least a vaccination scar, to pass 'quarantine'. In the early twentieth century, governed by a widely-endorsed eugenic policy (again by no means exclusive to Australia), British people had to undergo increasingly elaborate tests for syphilis, mental capacity, and alcoholism.[32] In the 1960s (at which point the annual emigration rate from Britain to Australia was around 43,000), when Australian incidence of tuberculosis was decreasing, it was assisted British migrants for whom a chest X-ray was compulsory.[33] In complicated ways, then, the structure of health screening which now in practice favors British entrants over entrants from virtually anywhere else on the globe, arose historically with them.

The UK has a different migration history. Obviously, but in this context importantly, it evaded altogether the Chinese diaspora/Chinese exclusion phenomenon of the late nineteenth and early twentieth centuries, which was the basis for health and disease sections of migration law in so many other national contexts (Australia, the US, New Zealand, Natal, Canada). Broadly, UK migration history involved Jewish migrants from Eastern Europe in the early 1900s; European Voluntary Workers in the 1940s; Irish migrants in the 1950s; migrants from the New Commonwealth in the 1960s and 1970s (Caribbean, India and Pakistan from the 1960s; those expelled from Kenya and Uganda in the 1970s); and asylum seekers and refugees more recently.

What was the relationship between the migration history of the UK and the development of health screening systems? It has been argued that the alleged relationship between anti-alienism and health exerted an influence on aspects of Victorian and Edwardian social reform.[34] However it is also striking that the mainstream secondary literature on the history of British immigration says remarkably little about health.[35] And in many ways rightly so, for there is scant evidence in the UK official record of the kinds of disease and migration links so evident in the Australian case. While the UK had a similar island-status to that of Australia (although continental Europe is geographically close), processes of quarantine were less rigid. The tests that were established were neither rigorous nor elaborate, and, as we have seen, screening for tuberculosis was conducted at the local level rather than at the border or offshore: the principle was one of conditional entry.

Nonetheless, there were certain links between screening and migration regulation in the UK, which need clarifying. The insertion of health clauses into the relevant legislation in the UK can be traced through the 1962 and 1968 Commonwealth Immigrants Acts, and through the 1971 Immigration Act. The available mechanisms for medical examinations for migrants were based on earlier public health legislation and migration regulations. In theory, medical examinations could be established, by amending section 143 of the 1936 Public Health Act whereby the Minister of Health could frame regulations 'for preventing danger to public health from vessels or aircraft arriving at any place'.[36] But while every alien seeking to land in the UK was liable under the 1953 Aliens Order to be medically examined at the Port of Entry, this had proved impracticable.

Given popular and some expert linking of immigration with disease, the 1962 Commonwealth Immigrants Act did specifically cite health grounds as a reason for refusal of entry. The Act stated that Immigration Officers could refuse entry into the United Kingdom of Commonwealth citizens 'if it appears to the Immigration Officer on the advice of a Medical Inspector ... that he [the migrant] is a person suffering from mental disorder, or that it is otherwise undesirable for medical reasons that he should be admitted'.[37] Moreover instructions issued to Medical Inspectors stated that the aims of medical examinations were to prevent the entry of, and bring to notice, any migrant who, if admitted, might endanger the health of other persons in the country; were unable to support themselves or their dependants in the UK; or who required major medical treatment. Immigration Officers could refer to the mentally and physically abnormal; those not in good health; the 'bodily dirty'; and those who mentioned health as a reason for their visit. A note added that physical conditions could include pulmonary tuberculosis, venereal disease, leprosy, and trachoma.[38]

Despite this legislation and the instructions issued to Medical Inspectors, the main emphasis at this time was on the forwarding of addresses of arriving migrants to the Medical Officers of Health (MOH) in the areas where the migrants planned to settle. The principle of conditional entry, that migrants were allowed to enter the country as long as they reported to a MOH, was clarified in the 1968 Commonwealth

Immigrants Act. A migrant could be refused admission to the UK or might be admitted to the UK by an Immigration Officer on condition that he reported his arrival to a MOH and attended for any tests or examinations required. This condition could only be imposed if, on the advice of a Medical Inspector or qualified medical practitioner, 'it appears to him [the Immigration Officer] necessary to do so in the interests of public health'.[39] The power to require medical examinations after entry was reaffirmed in the 1971 Immigration Act, where the Immigration Officer could give the migrant leave to enter the UK, but require him to report his arrival to a MOH and attend for any tests or examinations required.[40] This remains the legislation governing the screening systems employed at the current time of writing.

What is clear is that in the period up to 1970, relatively few migrants arriving at the Ports of Entry underwent chest X-rays, and this was even true of Heathrow Airport, the main Port of Entry, where chest X-rays had been set up on an experimental basis from February 1965. In 1969, for example, out of 44,575 Commonwealth migrants medically examined at Heathrow, only 4,229 were X-rayed (9 per cent).[41] Moreover in the 10-year period 1965–1975, of 402,583 New Commonwealth migrants 'seen' at the Health Control Unit at Heathrow, only 73,619 of these (18 per cent) had been X-rayed.[42] Therefore while substantial numbers of asylum seekers were referred for chest radiography at Heathrow in the 1990s, in the earlier period chest X-rays were a minor part of medical examinations. The establishment of the experimental scheme at Heathrow from February 1965 reflected the serious objections that had been mounted to compulsory chest X-rays at the Ports of Entry, both practical, and in terms of their scientific validity in detecting latent disease.

The principle of conditional entry in twentieth-century UK immigration law makes it the exceptional case internationally. As we have seen, nearly all immigration law from the pre-World War II period had health clauses, and indeed this was the case for much immigration law of new post-colonial nations in the post-war period.[43] UK history with respect to health screening begins to emerge as very different indeed to both settlement and post-colonial nations. At the same time, similar anxieties about human and microbial flow over borders, and the apparent need to distinguish between 'us' and 'them', were strongly evident in public and media fora in the UK, and increasingly so.

Who were 'the Others'?

Drawing borders around territory to produce 'us' and 'them' does not simply reflect divisions, but also helps to create hierarchized difference. As Augé remarks, it is intolerance and fear of the other that itself creates and structures nationalism and regionalism.[44] The conflation of others with ideas of contagion has been a common historical response. Yet the UK and Australian instances show how local and specific this response can be. Just who constituted the contagious 'other' has differed markedly in the two histories.

In the 1950s, the alleged high incidence of tuberculosis among migrants led to an outcry from the British Medical Association which accused the Ministry of Health of failing to implement a system of compulsory medical examinations, and noted that measures to guard against tuberculosis meant that Britain lagged behind Canada, South Africa, and the USA. This emphasis on the importation of disease was echoed in press coverage. Whereas migrants entering Australia and New Zealand were examined in their own countries, and some Australian states compelled migrants to be X-rayed within a month of arrival, the *Daily Herald* argued in February 1953 that 'Britain keeps no such close watch on immigrants'.[45] Medical examinations for aliens did not include chest X-rays, and only parties of European Voluntary Workers had to produce certificates of freedom from tuberculosis when they entered the country. Similar post-war stories about tuberculosis appeared in national, regional, and local newspapers. Concern was expressed that migrants were filling clinics; depriving British citizens of hospital beds; and that a disease previously thought to be beaten had re-emerged. The focus was largely on the costs of treatment, although this was occasionally tempered with relief that few 'Britons' were affected.[46]

The post-World War II struggle over where and how to undertake medical examinations suggests a desire to establish a sure and clear boundary at ports and airports.[47] Experts were also concerned. In June 1962, for example, the *Daily Express* reported the Deputy Medical Officer of Health at Dover as resigning on the eve of the implementation of the 1962 Commonwealth Immigrants Act, arguing of migrants that 'they can come in riddled with disease and remain undiscovered because it is quite impossible for the Medical Officer to examine them really adequately'.[48] As early as January 1962 an MP asked the Parliamentary Secretary to the Ministry of Health 'is there any possibility of immigrants being X-rayed before leaving their own country – as is apparently done in Australia?'[49] It is only more recently that this has become a real policy option.

Representations of otherness are also vividly apparent in correspondence between constituents, Members of Parliament, and Ministers of Health. In August 1960, for example, a Mrs Mitchell from Hove, in Sussex, wrote to Enoch Powell, Minister for Health, 1960–1963. The letter is worth reproducing in full. She wrote:

> People are becoming more & more alarmed by the reports that TB is again rearing its ugly head in our midst – due largely to the fact that it is being brought into the country & spread around by Commonwealth & Irish immigrants. We mothers of young teenagers (the most vulnerable

> group) feel that the Government (for whom I personally voted) is letting us down very badly by admitting these people. It just is not fair to re-introduce this horrible disease, when we have fought so hard over the years to raise our standard of living & health & were just beginning to reap the benefit, & to see this scourge almost wiped out. Please press for some sort of a check on these people – after all, Ireland isn't even in the Commonwealth, so why should we allow her sick people to contaminate & infect our society. The British public is pretty long-suffering, but one day – what with flooding it with coloured people, & now these horrible diseases – the worm may well turn. Please make a clean sweep.[50]

Mrs Mitchell's letter reflects the view that the most consistent migrant 'other' in British popular imaginings has historically been the Irish.[51] The irony is that for the post-war generation in Australia, the greatest risk was posed by British migrants themselves. Thus, while people in the United Kingdom anxiously watched migrant entry from Ireland, the West Indies, India and Pakistan, counterparts in Australia had suspicious eyes fixed on the likes of Mrs Mitchell herself.[52] Concerned citizens, especially in Western Australia which was the first port of call for ships from Britain, formed strong community anti-tuberculosis lobbies. They wrote in protest to Parliamentary representatives, Prime Ministers, and Ministers for Immigration, seeking increasing regulation of that well-known risk, the migrating Briton. The Anti-Tuberculosis Association of Western Australia noted in August 1947 that 'if he or she cannot come here with a certificate of freedom from Tuberculosis, then he or she is not a welcome addition to our population. No migrants at all would be better than migrants bringing Tubercular trouble'.[53] The Association and various senators successfully pressed the then Minister for Immigration, Arthur Callwell, to appoint a dedicated tuberculosis specialist in London, and to extend X-ray facilities under Australian authority, across the UK. Part of the problem, as Western Australian citizens saw it, was the discriminating regulations for chest X-ray: while British people on 'assisted' and 'free' passages (that is, the poorer) were compulsorily X-rayed, full-fare passengers (that is, the wealthier) were not.[54] It was understood that the broad social services available in Australia, especially under the 1949 Tuberculosis Act, would positively attract British 'consumptives' to those shores. This was one expression of a long tradition of British travel and migration to both Australia and New Zealand by tuberculosis sufferers, where the journey and the different climate itself were understood to be therapeutic.[55] Yet, the declining incidence of tuberculosis in Australia compared to Britain meant that it was British migrants who were commonly regarded (both popularly and officially) as 'the immigrant menace' until the mid-1970s. In terms quite unusual in Australian history (where 'British' has historically signified not difference, but identity), British people themselves were cast as the dangerous other: screening was necessary, not to keep out British people as a whole, but to keep out infected undesirables.

This changed rapidly from the mid 1970s when increasing numbers of refugees from South-East Asian crises sought asylum in Australia, often arriving by boat. Almost immediately, tuberculosis scares shifted from waning concern about British sources, to concern about Asian immigration and the incidence of tuberculosis in those countries. By that time, and because of the success in Western countries of antibiotic treatment, the global distribution of tuberculosis had radically altered. In Australia, this coincidence of newly low prevalence, and suddenly changed migration patterns from countries of high prevalence, had several ramifications. First, tuberculosis became discursively racialized in the way that leprosy and smallpox had been in earlier eras. This occurred in the printed press, in Parliamentary debate, on talkback radio, and in medical and expert discussion.[56] Second, successive Australian governments stepped up law and regulation with respect to migration and health: in particular the current stringent visa system was one result. Third, from 1989 so-called 'illegal entrants' were compulsorily detained in new detention centers, with public health functioning as part of the rationale for their mandatory segregation. Tuberculosis was back in the news and South-East Asian refugees were firmly identified as the main threat. For example, an article in the *Sydney Morning Herald* related how 'a disturbing number of refugees examined at Australian clinics have tuberculosis' and that the prevalence of tuberculosis amongst refugees was some 220 times that of the Australian population.[57] During the late 1990s, the anti-immigration stance of the Queensland Senator Pauline Hanson was widened to include health concerns, including tuberculosis. In March 1998 she stated that 'the immigration policies of Australian governments in the past two decades had allowed diseases such as hepatitis and tuberculosis to proliferate in this country'.[58] Recent changes towards stricter health screening in the UK suggests that such views are finally making their way into British policy and practice, if not law, as well.

Conclusion: Where is the border?

The border is traditionally the site where otherness is constructed and managed. Much of the recent work on border geographies has highlighted how the control of national borders, of belonging and not belonging, and of insider and outsider are experienced at different scales and in different places, not just at the boundary of the national state.[59] The control of borders has also moved into virtual space. In Australia, for example, there is an emerging 'cyberborder' where internet-based immigration control can process visa applications, issuing paperless visas to privileged applicants. Aside from challenging assumptions that the world can be neatly divided into discrete units of sovereign states, recent work offers new conceptualizations of borders as places of feedback, exchange and process, fluid rather than fixed. Donna Haraway, for example, argues that 'we are responsible for boundaries; we are they'.[60]

Comparing past and present practice of Australian and UK medico-legal border control allows us to reconfigure these histories and highlight the changing placement and nature of borders. Put simply, borders have been placed differently in these histories. W.F. Bynum has suggested that 'no man nor his nation is an island' with respect to infectious disease. But in fact the island-status of Australia and its geopolitical placement far from the 'Old World', adjacent to Asia, and as a 'white settlement', has been crucial both for the actual microbial and epidemiological history of the continent, and for public health management as it played out at the symbolic and cultural level.[61] The contiguous maritime border for Australia – the strong imagining of the nation as an island, surrounded by sea – created a historic significance for the literal edges of the geographic nation, enforced through maritime quarantine, a 'net' and network of quarantine stations, and the ongoing prominence of quarantine measures.

The key difference between Australian practice and UK practice has been the former's extension of the 'border process' well outside the national territory itself, historically to clinical and bureaucratic offices in the UK. Thus, while quarantine stations were frequently used for disinfecting and isolating infected ships and aircraft, and indeed for accommodating 'boat people' from Laos and Cambodia, there has never been an Australian equivalent to Ellis Island on the East Coast or Angel Island on the West Coast of the US: the equivalent was, more accurately, Australia House on the Strand which processed applications for travel and emigration. Thus the border of the potential Australian polity was and still is well beyond the territory itself – in country of origin, and 'offshore'. Indeed, the very terminology 'offshore' itself indicates how important 'the shore' is in the Australian national imaginary.

If the Australian health border is external to the nation's territory, the UK border is internal. The literal and symbolic site of the border in the UK instance is currently much more dispersed, and in many ways more ambiguous – certainly more porous in practice. Thus for many would-be migrants, breaching the geopolitical border leads to a controlled 'third space' neither inside nor outside the nation-state. This reflects both a different migration history and emerging globalization and transnationalism (for example, the impact of new trading blocks such as the European Union, particularly post-accession countries). As we have seen, the policy stance that was adopted and for the moment persists, that is, screening at ultimate destination, means incorporation of diseased people into the community. This is a deeply internal and dispersed border. Or perhaps this system is not usefully conceptualized in terms of borders at all. Rather, health is rendered a responsibility on the part of the entrant as the condition of entry and citizenship, rather like the US-Chinese in Shah's historical study: certain entrants may be territorially inside the nation-state, but belong only tentatively to the civic body.[62] That is, because screening is about both exclusion and inclusion, it is not just a spatial question, but a question of civic placement, where entrants are sometimes fully, but often only marginally located inside a new society. Health practices can reinscribe an ongoing marginal place, and this may be part of the significance of the diffuse UK system.

However, it is clear that this 'internal' system of UK management of disease and migration sits uncomfortably with the significance of national border security in the twenty-first century. There are signs of increasing confluence between the Australian and UK models of tuberculosis screening. This emerging shift in policy is about changing perceptions of disease, changing global distribution of disease, late twentieth-century global anxieties about human flow, and a transfer from the explicit racism of the pre-World War II era, with regards to health and migration, to an implicit racism within an officially 'race-less' policy. Whereas Australia has seen a move from the construction of British to South-East Asian and Middle-Eastern migrants as 'other' as it moves towards forms of post-colonial multiculturalism, the experience of the UK has been a move from the pathologizing of Irish, later Indian and Pakistani migrants, to asylum seekers and refugees more generally.

This chapter has attempted to contribute to broader debates about nationalism, migration, and public health by exploring the history of tuberculosis screening in the UK and Australia. In the UK instance, the health/migration border appears to be both strengthening and moving geographically outwards. The question is whether the new rigidity of policy in the UK may be a projection of desired migrant border control through a public health rationale, or if health screening is essentially now a legal way of displaying a firm border control that is not possible in the migration arena. Current UK changes toward a much more hard-line health screening at point of departure, possibly counteract the EU diminution of meaningful British borders. Health concerns could be seen as offering a convenient smokescreen to maintain control of the UK's borders in an age of both regionalization and globalization.

Acknowledgments

We would like to thank Richard Coker for much helpful advice, and the participants at a workshop held at the London School of Hygiene and Tropical Medicine in January 2005. Research was made possible by a grant awarded by the British Academy, the Australian Academy of the Humanities, and the Academy of the Social Sciences in Australia. Our thanks to Bernadette Power for research assistance.

Notes

1 A. Bashford and C. Strange, 'Asylum-Seekers and National Histories of Detention', *Australian Journal of Politics and History*, 48 (2002): 509–27; A. Bashford, 'The Great White Plague Turns Alien: Tuberculosis and Immigration in Australia, 1901–2001' in M. Worboys and F. Condrau (eds) *From Urban Penalty to Global Emergency: Tuberculosis Then and Now* (Montreal: McGill-Queens University Press, forthcoming); A. Bashford and S. Howard, 'Immigration and Health: law and regulation in Australia, 1901–1958', *Health and History*, 6 (2004): 97–112; A. Bashford and B. Power, 'Immigration and Health, Part 2: Law and Regulation in Australia, 1958–2004', *Health and History*, 7 (2005): 86–101.

2 A.M. Kraut, *Silent Travelers: Germs, Genes and the 'Immigrant Menace'* (Baltimore: Johns Hopkins University Press, 1994).

3 N.B. King, 'Immigration, Race and Geographies of Difference in the Tuberculosis Pandemic', in M. Gandy and A. Zumla (eds), *The Return of the White Plague: Global Poverty and the 'New' Tuberculosis* (London: Verso, 2003), pp. 39–54.

4 A.L. Fairchild, *Science at the Borders: Immigrant Medical Inspection and the Shaping of the Modern Industrial Labor Force* (Baltimore: Johns Hopkins University Press, 2003), p. 16.

5 In the official record this is called 'Port of Arrival' but to avoid confusion in the term 'Port', we use the term 'local destination system' to distinguish from the 'Port of Entry' screening in the US case, or 'offshore screening' in the Australian case.

6 J. Welshman, 'Tuberculosis and Ethnicity in England and Wales, 1950–70', *Sociology of Health & Illness*, 26 (2000): 858–82; J. Welshman and A. Bashford, 'Tuberculosis, Migration, and Medical Examination: Lessons from History', *Journal of Epidemiology and Community Health*, 60 (2006): 282–4; J. Welshman, 'Importation, Deprivation, and Susceptibility: Tuberculosis Narratives in Postwar Britain', in M. Worboys and F. Condrau (eds), *From Urban Penalty to Global Emergency: Tuberculosis Then and Now* (Montreal: McGill-Queens University Press, forthcoming); J. Welshman, 'Compulsion, Localism, and Pragmatism: The Micro-Politics of Tuberculosis Screening in the United Kingdom, 1950–65', *Social History of Medicine*, 16 (2006): 295–312.

7 Commonwealth of Australia, *Migration Act 1958*, section 10.

8 Schedule 1, Statutory Rules 2002, No. 285. Currently, the prescribed diseases are cholera, dengue fever, influenza, malaria, measles, polio, plague, rabies, SARS, smallpox, tuberculosis, typhoid fever, viral hemorrhagic fevers of humans, and yellow fever. Regulation 6, Quarantine Regulations 2000 (Cth); consolidation as of 26 March 2004. Note the recent addition of SARS and return of smallpox to this list.

9 Department of Immigration, Multiculturalism and Indigenous Affairs (DIMIA) 2002, *Procedures Advice Manual 3*, DIMIA, Canberra.

10 *Ibid.*

11 R. Mawani, 'Screening out Diseased Bodies: Immigration, Mandatory HIV Testing, and the Making of a Healthy Canada', chapter 8.

12 J. Jacobs, '(Post)Colonial Spaces' in M.J. Dear and S. Flusty (eds) *The Spaces of Postmodernity* (Oxford: Blackwell, 2002).

13 C.A. Van den Bosch and J.A. Roberts, 'Tuberculosis Screening of New Entrants: How Can it be Made More Effective?', *Journal of Public Health Medicine*, 22 (2000): 220–3; G.H. Bothamley, J.P. Rowan, C.J. Griffiths, M. Beeks, M. McDonald, E. Beasley, C.A. Van den Bosch and G. Feder, 'Screening for Tuberculosis: The

Port of Arrival Scheme Compared with Screening in General Practice and the Homeless', *Thorax*, 57 (2002): 45–9.

14 Joint Tuberculosis Committee of the British Thoracic Society, 'Control and Prevention of Tuberculosis in the United Kingdom: Code of Practice 2000', *Thorax*, 55 (2000): 887–901.

15 M.E.J. Callister, J. Barringer, S.T. Thanabalasingam, R. Gair and R.N. Davidson, 'Pulmonary Tuberculosis Among Political Asylum Seekers Screened at Heathrow Airport, London, 1995–9', *Thorax*, 57 (2002): 152–6.

16 R.M. Hardie and J.M. Watson, 'Screening Migrants at Risk of Tuberculosis', *British Medical Journal*, 307 (1993): 1539–40.

17 L.P. Ormerod, 'Is New Immigrant Screening for Tuberculosis Still Worthwhile?' *Journal of Infection*, 37 (1998): 39–40.

18 H. Hogan, R. Coker, A. Gordon, M. Meltzer and H. Pickles, 'Screening of New Entrants for Tuberculosis: Responses to Port Notifications', *Journal of Public Health*, 27 (2005): 192–5.

19 S. Millership and A. Cummins, 'Identification of Tuberculosis Cases by Port Health Screening in Essex 1997–2003', *Journal of Public Health*, 27 (2005): 196–8.

20 R. Coker, *Migration, Public Health and Compulsory Screening for TB and HIV, Asylum and Migration Working Paper Series* (London: Institute for Public Policy Research, 2003); R. Coker, 'Compulsory Screening of Immigrants for Tuberculosis and HIV: Is not based on Adequate Evidence and has Practical and Ethical Problems', *British Medical Journal*, 328 (2004): 298–300.

21 Conservative Research Department, *Before it's Too Late: A New Agenda for Public Health* (London, 2003), pp. 14–16.

22 B. Russell and N. Morris, 'Test Migrants for HIV and TB, say Tories', *The Independent* (15 February 2005).

23 'The Conservatives Propose a Radical Overhaul of the Immigration System', *The Economist* (27 January 2005).

24 HM Government, *Controlling our Borders: Making Migration work for Britain: Five Year Strategy for Asylum and Immigration* (Cm 6472) (London, 2005), p. 26.

25 G. Watters, 'The S.S. Ocean: Dealing with Boat People in the 1880s', *Australian Historical Studies*, 120 (2002): 331–41.

26 This history is detailed in chapters 5 and 6 of A. Bashford, *Imperial Hygiene: A Critical History of Colonialism, Nationalism and Public Health* (London: Palgrave, 2004).

27 Bashford, *Imperial Hygiene*, chapter 5; N. Shah, *Contagious Divides: Epidemics and Race in San Francisco's Chinatown* (Berkeley: University of California Press, 2001); R. Mawani, '"The Island of the Unclean": Race, Colonialism, and "Chinese Leprosy" in British Columbia, 1891–1924', *Journal of Law, Social Justice and Global Development*, 1 (2003): 1–21.

28 D. Walker, *Anxious Nation: Australia and the Rise of Asia, 1850–1939* (Brisbane: University of Queensland Press, 1999).

29 Bashford and Strange, 'Asylum Seekers and National Histories of Detention'.

30 M. Roe, *Australia, Britain and Migration: A Study of Desperate Hopes, 1915–1940* (Cambridge: Cambridge University Press, 1995).

31 S. Constantine, 'Empire Migration and Social Reform 1880–1950', in C.G. Pooley and I.D. Whyte (eds), *Migrants, Emigrants and Immigrants: A Social History of Migration* (London: Routledge, 1991), pp. 62–83; K. Paul, *Whitewashing Britain: Race and Citizenship in the Postwar Era* (Ithaca: Cornell University Press, 1997), pp. 35–63; S. Constantine, 'Waving Goodbye? Australia, Assisted Passages, and

the Empire and Commonwealth Settlement Acts, 1945–72', *Journal of Imperial and Commonwealth History*, 26 (1998): 176–95.

32 See 'Prohibited Immigrants. By One of Them', typescript in JHL Cumpston Papers, National Library of Australia, MS613 Box 7, c. 1920; 'O'Sullivan J – deportation due to tuberculosis' National Archives of Australia, Canberra A1 (A1/15) 1928/3804; Commonwealth of Australia, *Hansards*, House of Representatives, 5 October 1927.

33 Commonwealth of Australia, *Hansards*, House of Representatives, 26 October 1960; A.J. Proust, 'The Australian Screening Program for Tuberculosis in Prospective Migrants', *The Medical Journal of Australia* (13 July 1974): 35–7.

34 B. Harris, 'Anti-alienism, Health and Social Reform in late Victorian and Edwardian Britain', *Patterns of Prejudice*, 31 (1997): 3–34.

35 C. Holmes, *John Bull's Island: Immigration & British Society, 1871–1971* (London: Macmillan, 1988); C. Holmes, *A Tolerant Country? Immigrants, Refugees and Minorities in Britain* (London: Faber, 1991); Z. Layton-Henry, *The Politics of Immigration: Immigration, 'Race' and 'Race' Relations in Post-war Britain* (Oxford: Oxford University Press, 1992); W. Webster, *Imagining Home: Gender, 'Race' and National Identity, 1945–64* (London: Routledge, 1998). See also L. Marks and M. Worboys (eds), *Migrants, Minorities and Health: Historical and Contemporary Studies* (London: Routledge, 1997).

36 Public General Acts and Measures, *26 Geo. V and 1 Edw. VIII 1935–36, Vol. II, Public Health Act, 1936* (London, 1936), Chapter 49, p. 1320.

37 Public General Acts and Measures, *10 & 11 Eliz I and II Eliz, Commonwealth Immigrants Act, 1962* (London, 1963), Chapter 21, p. 115.

38 'Medical Examination of Immigrants Under the Commonwealth Immigrants Act 1962: Instructions to Medical Inspectors, 1962', National Archives, Kew, London [hereafter, NA] MH 55/2631.

39 Public General Acts and Church Assembly Measures 1968, *1968 Part 1, Commonwealth Immigrants Act*, 1968 (London, 1969), Chapter 9, p. 170.

40 Public General Acts and Church Assembly Measures 1971, *Elizabeth II, Part II, Immigration Act, 1971* (London, 1972), Chapter 77, p. 1696.

41 DHSS, *On the State of the Public Health, 1969* (London, 1970), p. 75.

42 M. Khogali, 'Tuberculosis Among Immigrants in the United Kingdom: The Role of Occupational Health Services', *Journal of Epidemiology and Community Health*, 33 (1979): 134–7.

43 A. Schloenharrdt, 'Exclusion of Infected Persons under Immigration Laws in Asia'. Paper presented to the Infectious Diseases and Human Flows in Asia workshop, University of Hong Kong, June 2005.

44 M. Augé, *A Sense of the Other* (Palo Alto: Stanford University Press, 1998).

45 'TB Aliens fill our Clinics', *Daily Herald*, 10 February 1953; 'Aliens with TB', *Evening Standard*, 11 February 1953; 'TB Being Imported: Preventive Work Set at Naught', *Surrey Comet*, 14 February 1953, NA MH 55/2275.

46 *Ibid.*; 'Foreign TB Cases are Depriving us of Hospital Beds', *Birmingham Gazette*, 6 February 1956; 'TB Alarms the Hospitals', *News Chronicle*, 5 August 1960; 'Scandal of Sick Immigrants', *Daily Mail*, 12 September 1961, NA MH 55/2277.

47 For detail on this see Welshman, 'Compulsion, Localism, and Pragmatism'.

48 'Health Check "Farce": Port MO Quits over Immigrants', *Daily Express*, 18 June 1962, NA MH 55/2633. In the same file see also 'Chaos at the Docks: Only Three Doctors at Airport', *Reynolds News*, 24 June 1962.

49 Graeme Finlay MP to Edith Pitt, 4 January 1962, NA MH 55/2633.

50 B. M. Mitchell to J.E. Powell, 5 August 1960, NA MH 55/2277.
51 L. McDowell, 'Workers, Migrants, Aliens or Citizens? State Constructions and Discourses of Identity Amongst Post-War European Labour Migrants in Britain', *Political Geography*, 22 (2003): 863–86.
52 See Welshman, 'Compulsion, Localism, and Pragmatism'.
53 A.J. Bishop, Secretary, Anti-Tuberculosis Association of Western Australia to N. Lemon, MHR, 1 August 1947, National Archives of Australia, Canberra, A436 (A436/1) 1950/5/2814.
54 Correspondence between Anti-Tuberculosis Association, Minister Callwell and Minister Holt, 1946–1950, National Archives of Australia, Canberra, A436 (A436/1).
55 L. Bryder, '"A Health Resort for Consumptives": Tuberculosis and Immigration to New Zealand, 1880–1914', *Medical History*, 40 (1996): 543–71; J.M. Powell, 'Medical Promotion and the Consumptive Immigrant to Australia', *Geographical Review*, 63 (1973): 449–76.
56 This is detailed in Bashford, 'The Great White Plague Turns Alien'.
57 'Tuberculosis Among Refugees in Australia Causes Alarm', *Sydney Morning Herald* (8 April 1987): 4.
58 'Hanson: They Spread Disease', *Sydney Morning Herald* (22 March 1998): 5. See also 'Australian Melting Pot Brought to Boiling-Point', *Guardian* (11 October 1996): 15; 'Hanson: Migrants Spread Disease', *Sydney Morning Herald* (2 July 1997): 4.
59 R.J. Johnston, D. Gregory, G. Pratt and M. Watts, *Human Geography* (London: Blackwell, 2000).
60 D. Haraway, 'A Manifesto for Cyborgs: Science, Technology and Social Feminism in the 1980s', in L. Nicholson (ed.), *Feminism/Postmodernism* (London: Routledge, 1990).
61 W.F. Bynum, 'Policing Hearts of Darkness: Aspects of the International Sanitary Conferences' *History and Philosophy of the Life Sciences*, 15 (1993): 422.
62 Shah, *Contagious Divides*.

7

Medical Humanitarianism in and Beyond France: Breaking Down or Patrolling Borders?

Miriam Ticktin

In 1998, France officially instituted a humanitarian clause in the law, granting legal permits to those in France with pathologies of life-threatening consequence, if they were declared unable to receive proper treatment in their home countries. This 'illness clause' was put in place to permit undocumented immigrants to receive treatment in France.[1] The logic behind this was humanitarian; the French state felt it could not deport people if such a deportation had consequences of exceptional gravity, such as their death. It was the lobbying of medical humanitarian groups such as *Médecins Sans Frontières* (MSF) or 'Doctors Without Borders' and *Médecins du Monde* (MDM) or 'Doctors of the World' that helped institute the illness clause as law in France.[2]

In this moment in French immigration history – and in particular, between 1997–2004 – those with cancer, HIV, polio or tuberculosis, or those who were understood as suffering from a medicalized form of trauma, became the most mobile, the most able to travel without hiding themselves in cold-storage containers or making mad dashes across the Channel Tunnel, risking their lives in the process. While I applaud the fact that those who are sick or disabled can indeed travel across borders into France, particularly when they are often prevented or discouraged from doing so in other national contexts studied in this book, the problem arises in allowing *only* such people to travel. In the absence of other immigration policies or means of legal access, the illness clause forces people who want papers to configure themselves as sick or suffering, and hence recognizable to humanitarian organizations, doctors or medical officials. Indeed, in this case, illness becomes the primary relationship between immigrants and the state: illness and/or suffering functions as a passport in the absence of other means of entry.

In a global political climate in which immigrants and refugees are increasingly labeled as criminal or economically burdensome, and in which the word 'immigration' is often paired with that of 'security', we must ask how the French state reconciled the denial of papers to immigrants because they were perceived to be criminal or economically burdensome, with the decision to give papers and social services to immigrants who were sick? Stated otherwise, why is it that illness is allowed to travel across borders, while poverty cannot? And in what ways is this situation specific to France?

This chapter examines the new medical humanitarianism,[3] suggesting that it constitutes the critical underlying context for the illness clause, and for the way in which sick bodies can travel across borders in this instance. Grounded in the French republican principles of universalism, which hold that all human beings are equal regardless of political, cultural, religious or other affiliation, medical humanitarianism offers the possibility of a world where borders are less relevant, of freedom and equality for all. It is built on the belief that humanity should not be divided into those who may live and those who must die.[4] Its goal is to save as many lives as possible. Yet is this goal, while universal in ambition, applicable across borders? In what sense is this type of humanitarianism limited to the French cultural and political conditions in which it was produced? And in what senses do these conditions determine its characteristics, even when it travels?

Elsewhere, I have analyzed the details of the illness clause itself.[5] Here, I discuss the genesis and characteristics of the humanitarian institutions that made it possible. Indeed, by addressing the case of medical humanitarianism, my larger goal is to examine one instance of 'universalism' in the contemporary world, how universal ideals are linked to national and colonial cultures, and how health and medicine help turn national ideals into transnational practices. I argue that while medical humanitarianism opens borders for some, it also inherits similar blind-spots of the universalism that grounded the French colonial 'civilizing mission'. Indeed, humanitarianism has its own tensions and exclusions, which often end up being expressed as a limited version of what it means to be human. With medical humanitarianism playing an increasingly important role in the contemporary world – recognized, for instance, by MSF's Nobel Peace Prize in 1999 – this chapter grapples with the need to understand how medical humanitarianism, in the form of NGOs or international institutions, works to manage borders – the openings, as well as the closures.

The suffering body

The humanitarian discourse of the late eighteenth and early nineteenth century functioned by publicizing the details of cruelty inflicted in individual bodies.[6] Indeed, as Thomas Laqueur writes in his exploration of nascent humanitarian sensibilities, 'humanitarian narrative relies upon the personal body, not only as the locus of pain but also as the common bond between those who suffer and those who would help'.[7] This focus on the individual suffering body has carried through into the new medical humanitarianism, but another dimension has been added: its apolitical status. As Liisa Malkki notes in the context of Hutu refugees living in Burundi and receiving humanitarian aid, 'wounds speak louder than words. Wounds are accepted as objective evidence, as more reliable sources of knowledge than the words of the people on whose bodies those wounds are found'.[8] Here, the political or historical circumstances of the refugees are ignored in order to focus on the individual body of the 'pure victim'.

The suffering body plays a critical role in the French illness clause. The clause was instituted in France when other doors to immigrants and refugees were closing. Indeed, an inverse correlation has been documented between the statistics on political asylum and permits for medical reasons: as the number of people admitted for political asylum decreased, those entering under the auspices of medical humanitarianism increased.[9] In this context, I have suggested that the illness clause was ratified only because it was perceived as outside the political realm.[10] That is, it was instituted in May 1998 as an amendment to the Ruling (Ordonnance) on the Conditions of Entry and Residence for foreigners, specifically, that which concerns the right to 'private and family life'. Placing this clause under the aegis of the 'private' exempts it from debates about the politics of immigration, citizenship, and notions of the French nation, and ignores the structural problems and economic demand that may have caused the immigration in the first place. Instead, the clause focuses attention on what is construed as an apolitical, suffering body. This clause is based on the notion of the universality of biological life[11] – a concept of life that overlooks all divisive political identities and affiliations – what Rorty refers to as 'species-membership'.[12] In my conversations with state officials and doctors, they confirmed that the space of pure life honored in the illness clause is conceived of in opposition to political community – in this sense, the boundaries of the political are demarcated from a purported universal, and hence legitimate, biological realm.[13]

These examples illustrate a notion of life as outside or beyond politics. Humanitarianism can be credited with legitimizing this notion of life in the contemporary world. Indeed, humanitarianism is grounded in the ethical and moral imperative to bring relief to those suffering and to save lives, regardless of political affiliation. It is precisely the ability to isolate victims in their present crisis, outside of politics and history – and to retain 'political neutrality'[14] – that allows humanitarian organizations to do their work, to render borders irrelevant in the name of a higher moral injunction to prevent and relieve the suffering of others. Yet, in what sense is this ability to move across borders in the name of political neutrality – and in the name of the suffering body – a French phenomenon, and in what sense does it have transnational articulations or possibilities? As chapters in this book detail, suffering in the form of HIV/AIDS does not travel across other national borders. There is a specificity to the French case, even as medical humanitarianism has contemporary transnational implications.

The new medical humanitarianism: a brief history

Humanitarianism has a long and sometimes ambiguous history, largely because it is not easily defined. Some call it a concern for the suffering of distant strangers;[15] others assert that it is framed by an ethics of intervention.[16] Religious orders dispensing charity and abolition efforts to ban slavery are early examples, but the direct precursor to the new medical humanitarianism – and an important player ever since – is the Red Cross movement, which was started in 1864 by a Swiss businessman, Henry Dunant, concerned by the plight of suffering soldiers. The mandate of the International Committee for the Red Cross (ICRC) was the protection of, and assistance to, victims of war. During and after World War II, a number of humanitarian organizations were born, such as Oxfam (Oxford Famine Relief Committee), some of which also fell under the rubric of the United Nations, such as UNICEF. French humanitarianism, however, was conceived considerably later.

Médecins Sans Frontières (MSF) was founded in 1971 by a small group of French physicians, headed by Bernard Kouchner. In the context of the civil war in Nigeria, these physicians served as volunteers in Biafra doing medical relief work among the Ibo. The founding of MSF grew directly out of Kouchner's reaction to the International Red Cross' strict approach to neutrality in war zones which involved a high degree of confidentiality because the emergency work of such international organizations was both made possible and constrained by agreements between states. The Red Cross and other international organizations had been providing equal amounts of relief to both sides in the Nigerian conflict, but in the spring of 1968 the government of Nigeria withdrew support for relief in an attempt to force the rebels into negotiations. A few NGOs like Oxfam decided to break with the tradition and airlift in food without the government's permission.[17]

Kouchner was one of the volunteers for the Red Cross who decided to flout the ICRC's policy on confidentiality, and talk to journalists about what they had seen. He stated: 'By keeping silent, we doctors were accomplices in the systematic massacre of a population'.[18] Returning to France from Biafra in 1969, he started an International Committee against genocide in Biafra, and in 1971, this group was constituted as a new species of NGO called 'Médecins sans Frontières'.

One of MSF's guiding principles is the desire to bear witness to the violations of human rights and human dignity that doctors encounter in the field. This resolve grew out of the finding that the International Committee of the Red Cross did not speak out against atrocities taking place in the camps in World War II, despite witnessing them through their deliveries of food and medicine: the Red Cross responded that their silence enabled them to continue doing relief work in the holocaust setting.[19] Kouchner and MSF's real innovation came in the way that they ruptured the silence: they informed a vast world-scale public about human rights abuses, and evoked public indignation about them, making sophisticated use of the mass media. They called this a 'duty to bear witness'.

Three defining features of MSF have fueled medical humanitarianism as a movement. As I will argue, they lay the groundwork for the new type of ethics exemplified in the illness clause, and the ways that French nationalism has been transformed into a form of internationalism,[20] which diminishes the significance of national borders. Let me make clear that MSF, perhaps more than any other NGO, is self-reflexive, constantly interrogating its role, its principles and its effectiveness. The following must be seen, therefore, as historically contingent features of the movement, rather than firm principles. First, MSF challenged and reconceptualized the notion of sovereignty. Second, they created the notion of the moral obligation to interfere in the name of suffering; and third, they strategically used the mass media to break down borders and bear witness to violations of human rights.

Their name itself – Médecins Sans Frontières – indicated a desire to put aside conventional borders of nation-states, to challenge sovereignty. While MSF never purported to suggest that borders are irrelevant – the name is about overcoming barriers more than borders[21] – they disavow any political or religious affiliation or identification, and assert their independence from political and governmental bodies. They do not agree that a nation should be free to determine its own destiny. Their vision is a global one governed by the principles of the Universal Declaration of Human Rights. Their vision of medicine is also transcendentally universalistic: they claim that because 'illness and injury do not respect borders', neither should medical care.[22] Despite the extant mandate instituted in 1948 by the Universal Declaration of Human Rights to put aside the rules of sovereignty in cases of human rights abuses, this mandate was never insisted upon, let alone acted upon in so clear a way. MSF in many ways spearheaded a group of what researcher James Rosenau has called 'sovereignty-free' actors that include Greenpeace and Amnesty International, each positioning themselves on an international stage previously reserved for states.[23]

This universalistic aspect of MSF has a distinctly French flavor; that is, the French have long grounded their politics on a paradigm of universalism. This is what undergirds French notions of Republicanism: a belief in equality and the denial of the relevance of particularities. In this same way, MSF emphasizes a universalistic conception of human worth, and the unity of the human condition, playing down different social and cultural attributes of both the doctors and those whom they assist.[24] This universalism was clearly evident in French politics at the time. A member of the Communist Party and as a medical student, Bernard Kouchner was a child of the May 1968 student movement in France. Indeed, the French medical sociologist Claudine Herzlich argues that medical humanitarianism rose out of the context of 1968, where the School of Medicine in Paris was a place of high activism. The medical students drafted a 'white book' containing a radical reform of medicine, medical education and the practice of medicine, trying to make a break with 'capitalistic medicine'.[25] Herzlich argues that Kouchner is in many ways representative of that movement and the fight it waged against capitalism. His will to practice a different medicine resulted from the failure of May 1968, and she explains that the doctors who were involved in '68 now practice a type of medicine linked to public campaigns and actions, and to a universal cause whose symbolic value reaches well beyond the field of medicine. Kouchner's universalism draws on the tradition of the 'Rights of Man' that is conceptualized as inherent to French civilization, and linked to French nationalism. MSF works on the premise that this tradition must be upheld at all costs. In other words, wherever suffering exists, it must be alleviated: all people have an equal right to exist without violence or pain. Indeed, there have been a multitude of organizations both professional and voluntary born from this idea of 'sans-frontièrism'; there are now groups that range from 'reporters without borders', and 'engineers without borders' to 'chiropractors without borders'.

The second defining feature of MSF, *le droit d'ingérence* or *the right to interfere*, is based on the universal conception of the human condition; sovereignty is challenged in the name of this universal humanity. This is a particularly interesting idea, because *le droit d'ingérence* can be translated into English in several different ways: it can mean the 'right to *intervene*', but it can also signify the 'right to *interfere*', and the latter translation is preferred by Kouchner. In fact, the preferred translation is the '*duty* to interfere', which again plays with the word 'droit'. In English, 'droit' could signify the '*law* to interfere' but in French, the preferred meaning has a connotation of moral obligation. In English, the idea of a 'duty' carries more weight than that of a 'right'. Yet, as Allen and Styan point out, Kouchner takes the *duty* to interfere as a given – one is morally obligated to interfere if others are suffering. As a result, he has spent his time fighting to institute the legal *right* to interfere.[26] Again, this draws from a French political and legal tradition: the French practice of international law regarded intervention as technically legal, while others interpreted it as infringing on the rules of sovereignty.[27] In this sense, MSF suggested that 'ingérence' be more actively developed in the wider international community.

Early in MSF's life – 1979 – there was split between its founding members and members of the second generation over the extent of intervention or interference, and where the line should be drawn. The split resulted from a disagreement about Vietnamese 'boat-people' who were dying in the sea by the thousands. Kouchner wanted MSF to invest its resources to charter a boat to rescue these refugees, and he joined with prominent French intellectuals to mobilize the media and draw support for the cause. The younger members of MSF felt that it surpassed their competence and the capacities of medical humanitarianism, and voted against him. Rony Brauman, the president of MSF from 1982–1994, was amongst this younger group, and writes that the disagreement was about whether to forge an independent organization of doctors, with specific guidelines and limits (later called private humanitarianism), or a series of symbolic actions, that ultimately remained state-affiliated (state humanitarianism). This in fact remains an ongoing tension in the notion and practice of humanitarianism. There were accusations of betrayal on both sides, but ultimately Kouchner and many of the other founders left and formed *Médecins du Monde* (MDM) or Doctors of the World.[28] These remain the two most prominent medical humanitarian organizations, and for the most part share the same goals; nevertheless, the split highlights the ongoing debate and tension about how much 'interference' is enough, and how 'moral obligation' or 'right' is defined.[29]

Part of the debate over le *droit d'ingérence* involves France's colonial history, and subsequent post-colonial political developments. French colonial policies were highly interventionist, and were involved in nearly all aspects of life, from health to education to sexuality. With the war in Algeria and the independence movements of many French colonies in the 1950s and 60s, the New Left led by student youth espoused a Marxism that identified and sympathized with the international proletariat in the Third World. This movement was called *tiermondisme* or 'third-worldism' and took a position of solidarity with revolutionary movements in the 'Third World', rebuking any criticism of these independence movements, and any type of interference with their regimes: this was in reaction to the previous interventionist colonial policies. In the late 1970s and 80s, however, it became more apparent that the anti-colonial revolutionary Marxist movements were not necessarily successful and intervention again became an issue. As MSF president at the time, Rony Brauman claimed, 'Some 90 percent of refugees in the third world were fleeing this kind of regime, and we were working in camps where they had been gathered ... We were ferociously critical of third worldism and what [Raymond] Aron called the 'lyrical illusion' that it permitted'.[30] Specifically, a division emerged between those who reaffirmed the original anti-colonialist position (the original *tiermondistes*), and those who wanted to promote a new *mission civilatrice* or 'civilizing mission' in the name of human rights. MSF fell into this latter category, as did the broader 'without borders' movement.[31] In other words, a form of universalism based on rights explained the 'moral obligation' to once again interfere, walking in the footsteps of the colonial legacy, even as it worked to counter practices that resulted from centuries of colonialism and the struggle to decolonize. The 'need for defense of human rights in the "Third World"' was endorsed by President Mitterand in 1987, and his government included 'humanitarianism of the State' in its policies. Indeed, there was a place for the emerging humanitarianism in the French state itself: the Minister of Human Rights from 1986–1988 was Claude Malhuret, an ex-president of MSF. And in 1988, Kouchner was named French minister of health and *action humanitaire* – the latter being a position created especially for him by the newly-elected government.

This *droit d'ingérence* has been gradually instituted internationally: it can no longer be accurately labeled just 'French'. Kouchner likes to claim that this relates to his influence.[32] He has held political office in France and has been engaged with proposing and framing the plethora of UN resolutions that have created a new framework for international law. UN General Assembly Resolution 43/131, passed in 1988, broke new ground by stating that abandoning victims without humanitarian assistance 'constitutes a threat to life and an offense to human dignity', and stressed the importance of 'intergovernmental and non-governmental organizations working with strictly humanitarian motives'. While resolutions are not binding, Allen and Styan write that this was widely taken to mean that cross-border operations in war zones were now formally acceptable; and Resolution 45/100 made this support for 'humanitarian corridors' more explicit.[33]

While acknowledging that all international movements begin in specific national contexts, it is also the case that medical humanitarianism has moved beyond being simply a French tradition. There are 19 current national MSF sections, five of which are fully operational in directing independent missions. MSF is now a mobile transnational entity.[34] Yet even within the French section, not all French politicians and intellectuals support the right to interfere; Rony Brauman (former president of MSF) was highly critical of Kouchner's attempt to institute humanitarian intervention as a legal right and duty, suggesting that international jurisprudence can be used by states when they find it convenient or in their interest to interfere. Brauman pointed to the inevitable mix between military and humanitarian aims in a 'humanitarian invasion' (or a legal intervention supported by United Nations or national military forces to help with transportation or supplies) and suggested that mixing medical humanitarianism with military interventions runs counter to the idea of allowing NGOs access to victims, and to medical humanitarianism's purported vocation of healing. Indeed, there is evidence that relief activities reinforce war economies.[35]

The third element characteristic of MSF in particular and medical humanitarianism more broadly is the use of the media, which functions as a critical part of the duty to give testimony. The mediatization of missions was considered essential to the new humanitarianism; not only did it publicize the findings of doctors as part of moral duty to bear witness, but the publicity helped to secure funding and provide a degree of immunity from governments and other political interest groups hostile to their interventions. Media publicity crossed borders and built solidarity. MSF quickly learned that official funding as well as private donations were most effectively secured when international media covered acute suffering. Kouchner had an intuitive sense of the role of media in politics, and NGOs all over the world have since followed his lead. Early on, MSF became a key source representing the 'Third World'.[36]

The success of medical humanitarian organizations in harnessing the media and in professionalizing NGOs is problematic. Clearly, there is a fine line between bearing witness to suffering, and sensationalizing it, turning human disaster and tragedy into a spectacle that is at once distant and close. Through television, people have witnessed death, starvation, and mass violence on an unprecedented scale. This mediatization runs the risk of dehumanizing and numbing both the victims and spectators, rather than creating a sense of empathy for a common humanity.[37] Moreover, such mediatization of suffering can be harnessed in self-interest, to push certain causes and to compete for funds, recognition and prestige. Yet, as Tom Keenan has noted, mediatization in many ways is no longer a choice: 'one cannot understand, nor have a properly political relation to, invasions and war crimes, military operations and paramilitary atrocities – both of maximal importance for human rights campaigners – in the present and the future if we do not attend to the centrality of image production and management in them'.[38] Thus, while the French first demonstrated the good that can come of publicizing suffering, the moral dilemma of what to do with that testimony now extends far beyond France.[39]

Perhaps the most important tension in the debate over the duty to bear witness and to publicize suffering is how this affects the purported political neutrality of humanitarian NGOs. As I suggested, a belief in political neutrality is what allowed the illness clause to come into existence in France – it was perceived to be apolitical, or above politics. As Kouchner himself states: 'If you are humanitarian...this is not politics, you must be neutral, taking care of all'.[40] Indeed, I attended the 2005 Annual International General Assembly meeting of MSF, and the question of political neutrality as related to the duty to give testimony was foregrounded: to what end does MSF bear witness? To advocate for justice, for policy changes, for an end to the impunity of certain actors? This, the representatives realized, quickly ran into the dangerous terrain of political engagement and intervention, which MSF prides itself on avoiding. It should be noted that this is in direct contrast to the 'Anglo-Saxon' or British tradition of humanitarianism exemplified by CARE or Oxfam, who choose to combine humanitarianism with development, taking a more utilitarian approach.[41]

To solve this problem of descending into politics through witnessing, Rony Brauman advocated for the 'description' of what one witnesses as opposed to 'qualifying' it, which entails making a judgment. However, the line between these is not always so clear. For instance, a representative from MSF-Holland suggested that bearing witness to rape in the crisis in Sudan required that one also call for justice for the victims. He cautioned about the slippery-slope between political neutrality and irresponsibility. There is an ongoing debate, then, not only in France, but transnationally, about medical humanitarianism and the moral obligations it entails – a debate in which all sides are informed by French history and political traditions, but also one to which each humanitarian worker and each crisis brings its own cultural, political and historical location. As Redfield states, 'MSF responds with a defense of life that both recognizes and refuses politics. It forcefully claims an independent right to speak out and act without regard to considerations other than conscience, yet it never quite abandons neutrality in its insistence that final responsibility for alleviating suffering lies elsewhere'.[42] The place of *témoignage* itself is constantly debated, and the current MSF charter does not even mention *témoignage* in an attempt to avoid suspicion by local authorities that missions have an element of political intent. Instead, bearing witness is framed as a choice for members, rather than a moral duty, even as the guide suggests that the organization itself feels morally bound to speak out.[43]

The medical humanitarian NGOs have expanded in several directions since their founding. They have moved beyond emergencies to focus on longer-term projects such as MSF's 'Campaign for Access to Essential Medicines', which is a sustained effort to attend to the difficulties of poor people in obtaining medicines for conditions such as HIV/AIDS, malaria and tuberculosis. Similarly, many new medical humanitarian organizations have sprung up in Western locales with variations on the same theme: Physicians for Human Rights, Global Lawyers and Physicians for Human Rights, Physicians for Global Survival, to name but a few. These developments pose constant challenges to the conception of medical humanitarianism, adapting and re-adapting the French tradition: for instance, while the Access to Essential Medicines campaign grows out of frustration at drug shortages in MSF's work, it borders on playing an advocacy role in the realm of policy, again, something MSF has strictly refused, acknowledging that it cannot save the world by itself. The British humanitarian organizations see fewer problems with this combined approach, and their organizations tend to be more thematically than professionally organized. Thus, for instance, Oxfam is organized in opposition to poverty and hunger, which allows both emergency and long-term responses.[44] But in the transnational arena, we see a growing convergence between these two forms of humanitarianism.

Medical humanitarian organizations have also turned their attention to situations of socio-medical need in their own and other so-called 'developed' countries: they have brought their concerns back home, within their borders, and in ways which draw critical attention to national borders. With the effects of severe economic recessions in the late 1980s, MSF and MDM saw the need to help those excluded from systems of social security in France. In 1993 Bernard Granjon, the president of MDM-France, asked: 'Must we accept, in this rich country which is France, in the homeland of the rights of man and of the citizen, the ineluctable spiral of poverty which results in what more and more resembles professional and social apartheid?'[45] In France, these medical humanitarian organizations joined with other NGOs in 'Mission Solidarité France', a network of centers across France instituted to ensure that the socially excluded received free medical and social services, and further helped to reinsert the excluded into the mainstream social security system. With this same impetus to bring humanitarianism 'home', MSF and MDM began to advocate for the *sans papiers* (undocumented), and with the help of the larger Collective for the Rights of Sick Foreigners, the illness clause was pushed through.

Inclusions and exclusions: reviving universalism and colonialism

Humanitarianism has been associated with a revival of universalist discourse, with the opening of borders to compassion, but also to intervention; MSF in particular was seen to offer an alternative to increasingly tired internationalist ideals – it was part of a Left with a new vision.[46] In some senses, it offered a fresh way to express the universalist French values of liberty, equality and fraternity. MSF and the values it embodied were – perhaps paradoxically – both a way to create a new internationalism, and also a new international role for France. The founders ultimately believed that there was no better nation to incarnate the universal dynamic.

Yet, as many have shown, a universalist ideology also grounded the colonial enterprise; in other words, a system that celebrates its universalism and tolerance can also maintain structures of racial and economic exclusion. While medical humanitarianism need not be bound to the same fate, especially given its characteristic reflexivity, as Redfield suggests, 'a borderless world retains the ruins of earlier frontiers'.[47] This imbricated history highlights the national ideological inheritance of medical humanitarianism. I examine briefly some of the tensions within the colonial civilizing mission itself and the links with colonial humanitarianism. I then turn to the inheritance of this colonial legacy both in MSF and in the MSF-inspired 'illness clause'.

Drawing on universalist rhetoric of 1789 regarding the right of all people to basic freedoms, republican ideology inspired the French to take measures to liberate Africans from indigenous forms of oppression they believed to exist, which included not only forms of African slavery and 'feudalism', but also disease. French colonial regimes operated on the principle that the colonized had the potential not only to be emancipated, but to be assimilated as citizens of the Republic: everyone had the potential to be equal, and to be free.[48] The goal was to make people citizens when they were ready – when they were 'evolved' enough.

Yet as early as the 1790s, principles of universal inclusion were combined with practices of racial exclusion. It was the universalist ideal of the unity and fundamental equality of all humankind and its uniform capacity for civilization that provided the justification for colonial violence and oppression: in the name of equality, imperial forces constructed knowledge of non-Western cultures as inferior, in need of intervention and civilizing. Colonial subjects were seen to be unready and immature, but the 'generosity' of the French persisted in civilizing those '*evolués*' – those who were evolved.[49] Assimilation depended on a gradual process of transformation through education, from tradition to modernity. So, universal citizenship – while held up as goal and ideal – was for the most part deferred.

As Dubois notes, the idea that people were not ready for citizenship is one that was at the heart of the contradictory regimes of French emancipation, particularly in the French Caribbean in 1794 and 1848. Victor Hugues, for example, who abolished slavery in Guadeloupe in 1794, and later oversaw the re-establishment of slavery in Guyana in 1802, argued that the continued subjugation of the newly emancipated slaves could be justified by the fact that equality was not absolute, and that citizens had to be treated differently according to their moral and intellectual capacities. To learn how to be free, they had to be forced to work. As Dubois writes: 'Only then could the moral stain of generations of slavery be removed; only then could the ex-slaves become citizens'.[50] Dubois calls this deferral of the application of universal ideas 'Republican racism', and suggests, rightly so, that this continues to haunt contemporary discussions around immigration in France.[51]

Humanitarianism inherits the ideological underpinnings of republican racism, as I will discuss, in that those being saved must defer their political status in order to remain apolitical suffering bodies. It also inherits the contradictions of colonial humanitarianism. Humanitarianism itself expanded with colonial settlements, and there was a cadre of 'colonial humanitarians' who specifically worked to foreground the requirements of justice and morality, placing them on a par with the motives of interest. Importantly, Lester argues that colonial humanitarianism was not simply a legitimating screen for imperialism. It was formulated explicitly to challenge colonial military discourses and settler practices through a new global ethics – one which did not work to colonize land, but to 'colonise the mind. To bring down, as it were, the ideas and principles of Heaven... and to deposit them in understandings and hearts of the inhabitants; and thus to elevate and to save them'.[52] Interestingly, Lester also notes in the British context that the concern for distant strangers and pursuit of collective moral action served as a form of national pride – not unlike the way humanitarianism works today, intervening to clean up the mess of military intervention, while simultaneously affording a feeling of moral superiority.[53]

In addition to the practices of 'Republican racism' and colonial humanitarianism, MSF's genealogy includes a long line of colonial endeavors related to health, especially missionary activities. Albert Schweitzer's hospital in Lambarene figured centrally in colonial claims to a civilizing mission. Similarly, MSF inherits the special place occupied by Africa in both the French civilizing mission, and the representation of disease in empire. As Redfield points out, in 2002–2003 MSF spent over half its funds in Africa, keeping it the center of humanitarian activity.[54] A colonially-derived racial hierarchy inadvertently underlies their missions, where the doctor is still predominantly a White male, and often a European expatriate. In the name of the protection of human dignity, he speaks for and represents non-White, non-European victims. There is, then, a colonial legacy in humanitarianism generally and for MSF specifically, which recalls the exclusions and contradictions of universalism.

The humanitarian-inspired illness clause in current French law brings with it some of the same contradictions and blindspots of universalism. While we might have imagined that a concept of humanity based on the universality of biological life would bring us closer to equality, closer to a borderless world that overlooks all divisive political identities and affiliations, in fact, it encourages a limited and limiting version of who can travel across borders. Indeed, this universalism has resulted in a structural situation that favors sick and suffering bodies above all others.

In order to illustrate the limitations and colonial legacy of humanitarianism as exemplified in the illness clause, I turn to an insightful piece by medical doctor and anthropologist Didier Fassin about the social condition of HIV+ immigrants in France.[55] Fassin describes the case of a Nigerian man who initially found out he was HIV+ in his quest for legal documents in Germany. Unlike some other aspiring immigrants, this man neither infected himself nor deliberately configured his illness to get papers. Rather, the discovery of his positive status came as a result of the AIDS test in Germany, mandatory for his application for papers. He fled to France, and as Fassin astutely points out, his first experience of France was as an undocumented immigrant, with no home, not a sick person with HIV. He moved to France for papers, not for medical treatment. In other words, he experienced his positive status as a *political condition* in the sense that it expelled him from Germany. His condition was at once political, social and biological, but he understood it primarily through a political and social lens. After 4 years, however, this man did fall seriously ill. He was forced to undergo a surgical procedure, and after several months and many visits to social workers and associations, he was granted papers for 'humanitarian reasons', on the basis of the illness clause. Gradually, illness became the driving force of his life. It gave him legal papers, an apartment, and a minimum financial allocation, which he would not have otherwise had.[56] His biological condition determined his social condition – that is, insofar as he proved he was suffering and life-threateningly sick, he was granted social and legal status in France. Of course, because the illness clause is dependent on life-threatening conditions, it is in fact his physical life that he trades in for social recognition – the prospect of his death is what ensures his social life.

Fassin argues that AIDS is thus a social condition in France. While in most countries, HIV/AIDS is a dangerous secret that immigrants must keep from the state at all costs, in France, HIV/AIDS can act as the primary relationship between the immigrant and the state, actually bestowing legitimacy. However, in this realm where biological life is regarded as primary, and a politics of humanitarianism and benevolence come into play, special assistance or entitlement rights serve to segregate rather than to integrate. This politics of humanitarianism requires that people remain ill, that they remain exceptional. The renewal of papers is dependent upon their continued identity as victims. Thus, if granted a 1-year residency permit, one must renew this five times before being granted a 10-year residency permit, and only then is one eligible to apply for citizenship. During this time, one must inhabit the subject position of dependency. Indeed, entitlement rights, of which the illness clause is one example, mark subjects within structures of economic stratification and cultural or ideological valuation. Their distinguishing characteristic is dependence. As the history of poor relief or the establishment of public orphanages might imply, one need not have the right to vote to receive some form of public support, and vice versa: public assistance does not necessarily open the way to political rights or citizenship.[57] Subjects dependent on public provision for basic needs are marked as a particular type of subject, not quite equal, not quite able to take care of themselves.

This relation of dependence takes on added significance as it maintains power dynamics deriving from the colonial era: the majority of *sans papiers* in France come from former French colonies. In a neo-colonial relation, one is able only to be assisted or rescued, and – crucially important – this status is exceptional: one is by definition not considered 'normal'. In this sense, the *sans papiers* who receive papers through the illness clause are reproduced as physically vulnerable. They are maintained as a special and extraordinary category of subjects.

Here, we can see how medical humanitarianism in the form of the illness clause works to create subjects who will never be equal; they, like the subjects of colonial regimes, are only recognized through their lack. Just as colonial subjects could never become civilized enough, the subjects of humanitarianism forever miss the jump to full citizenship. To maintain their status as recognized subjects, they must remain either sick, suffering, or displaced – they must remain subjects of benevolence, not of full rights.

Humanitarianism and the 'anthropological minimum'

Is humanitarianism inherently flawed in its ethical goals? Uday Mehta helps to explain how these limitations might derive from the exclusions built into the notion of universalism, as embodied by liberalism, and in particular, liberal imperialism.[58] He suggests that the base standard of universal human nature described by Locke, what he calls 'the anthropological minimum' – that humans are equal, free and rational from birth – is in fact too minimal, and too devoid of context to be substantiated. In other words, it ignores the specific cultural and psychological conditions woven in as preconditions for the actualization of these capacities. It assumes certain characteristics are common to all human beings, with which they are born – it does not leave room for their development. It takes human beings out of all sociological or historical context, again in order to fix a universal set of characteristics about human nature. This anthropological minimum therefore allows for strategies of exclusion based on implicit divisions and exclusions in the social world. If one does not exhibit the expected characteristics of human nature – for instance, if one behaves in a way that is differently rational, and hence unrecognizable in liberalism's terms – liberal universalism locates this outside of the anthropological minimum, and hence, outside the category human. For example, Mehta explains how, in the British Empire, Indians were treated as inferior, and therefore governed without freedom, because their purported 'inscrutability' led to the belief that they were like children, lacking in appropriate rationality. Without rationality, according to the British interpretation of the anthropological minimum, one could not be counted as fully equal, or, for that matter, fully human. In other words, because Locke neglected to qualify the context of his concept of universal human nature, exclusions based on the different qualifications of 'human nature' were justified.

Humanitarianism protects a similar universal, but minimal and acontextual vision of life, in this case, defined by the capacity to suffer. As such, it also allows for the differential and unequal treatment of people who are not recognized within this minimum, historically contingent standard. If undocumented immigrants – or refugees or victims of war, for that matter – have their lives saved by humanitarian action, it is not clear what notion of life this entails. Does the life saved come with cultural attributes, political inclinations, linguistic skills, skin color? This is not made clear. These attributes, these qualifications, belong to the political realm. They are not part of the humanitarian mission, which responds to suffering in its most rudimentary and often biological forms, such as hunger and disease. As Redfield suggests, 'humanitarian action can preserve existence while deferring the very dignity or redemption is seeks'.[59] Indeed, by responding only to emergencies, limiting action to a temporal frame of the present and therefore deferring political solutions, humanitarianism can only defend a minimal existence. And, when different forms of suffering are not recognized by the conceptual framework employed by humanitarian workers, they are not responded to: that suffering is placed outside the minimal conditions requiring response, as in the case of Mehta's liberal universalism. Such people are thus similarly located outside the definition of human. In this form of minimalism, too, there is room for exclusions and hierarchies. Thus, the universalism of humanitarianism only works for a very basic notion of humanity; the rest is qualified and protected or stopped from crossing borders by the political context in which each person is found – which in our contemporary world, remains the nation-state.

Conclusion

Medical humanitarian organizations have largely reconfigured how we understand borders – crossing them in the name of a transnational moral imperative – and such organizations also help individuals to survive if and once they cross borders. However, these individuals can only cross – and remain on new territory – as specific types of limited subjects: in refugee camps, and in France, for instance, as sick or suffering bodies. This said, let me be clear: to blame humanitarianism for not solving the world's political problems and creating a just world is not my goal. Humanitarianism does not act with longer-term political consequences in mind. In fact, it avoids precisely this, in the name of immediate, urgent and temporary care, and in the name of political neutrality. Its constituents are clear that it will not – and should not – save the world. My goal, rather, has been to explore the ways that medical humanitarianism embodies a universalist ethic that enables border crossings, that enacts a notion of (apparent) equality. Insofar as medical humanitarian organizations ground their action on the desire to relieve suffering, and do not qualify this notion further, they allow for differential protection, and for inequality to flourish; in this sense, humanitarianism inherits, and remains limited by the paradoxes of French universalism on which it is founded. What travels across borders is a basic understanding of what suffering entails – defined differently in each context – while the human, qualified beyond suffering by its political, social, cultural, economic and religious dimensions – remains suspended at the border.

Notes

1 The law states, 'une carte de séjour temporaire est délivrée de plein droit à l'étranger résidant habituellement en France dont l'état de santé nécessite une prise en charge médicale dont le défaut pourrait entrainer pour lui des conséquences d'une exceptionnelle gravité, sous réserve qu'il ne puisse effectivement

bénéficier d'un traitement approprié dans le pays dont il est originaire'. Or, 'a temporary residency permit is granted to the resident foreigner in France whose state of health requires medical treatment in the absence of which there would be consequences of extreme gravity; this is subject to the foreigner's inability to obtain appropriate treatment in his/her country of origin' (author's translation).

2 MSF and MDM joined with other NGOs to form the 'Collective for the Rights of Sick Foreigners in France' which was comprised of 35 associations or NGOs including associations for sick people, doctors' organizations, trade unions and associations for immigrant rights.

3 I refer here to the humanitarianism that began with Médecins Sans Frontières in the early 1970s.

4 J.H. Bradol, 'Introduction: The Sacrificial International Order and Humanitarian Action' in F. Weissman (ed.) *The Shadow of 'Just Wars': Violence, Politics and Humanitarian Action* (Ithaca: Cornell University Press, 2004), p. 9.

5 M. Ticktin, 'Between Ethics and Politics: The Violence of Humanitarianism in France', *American Ethnologist*, 33 (2006): 33–49.

6 See A. Lester, 'Obtaining the 'due observance of justice': the geographies of colonial humanitarianism', *Environment and Planning D: Society and Space*, 20 (2002): 277–93.

7 T. Laqueur, 'Bodies, Details and Humanitarian Narrative' in L. Hunt (ed.) *The New Cultural History* (Berkeley: University of California Press, 1989), p. 177.

8 L. Malkki, 'Speechless Emissaries: Refugees, Humanitarianism and Dehistoricization', *Cultural Anthropology*, 11 (1996): 384.

9 D. Fassin, 'The biopolitics of otherness: Undocumented foreigners and racial discrimination in French public debate', *Anthropology Today*, 17 (2001): 3–7; D. Fassin, 'Une double peine: la condition sociale des immigrés malades du sida', *l'Homme*, 160 (2001).

10 Ticktin, 'Between Ethics and Politics'.

11 I draw partly on Giorgio Agamben's notion of 'bare life', based on the Greek *zoe*, or life devoid of political and social qualification. G. Agamben, *Homo Sacer: Sovereign Power and Bare Life*, D. Heller-Roazen (transl) (Stanford: Stanford University Press, 1998).

12 See R. Rorty, 'Human Rights, Rationality and Sentimentality', *Yale Review*, 81 (1993): 1–20.

13 As Didier Fassin suggests, discussing immigration in France, 'the legitimacy of the suffering body proposed in the name of a common humanity is opposed to the illegitimacy of the racialized body, promulgated in the name of insurmountable difference'. Fassin, 'The biopolitics of otherness', 4.

14 This is a term I will discuss later in the essay, but it should be noted it is highly contested what this neutrality involves, and the ICRC model of neutrality is different from that of MSF.

15 For more on this perspective, see L. Boltanski, *Distant Suffering: Morality, Media and Politics*, G. Burchell (transl) (Cambridge: Cambridge University Press, 1999); T. Haskell, 'Capitalism and the Origins of Humanitarian Sensibility' in T. Bender (ed.) *The Antislavery Debate: Capitalism and Abolitionism as a Problem in Historical Interpretation* (Berkeley: University of California Press, 1985); Lester, 'Obtaining the due observance of justice', 277–93.

16 I. Feldman, 'The Quaker Way: Ethical Labor and Humanitarian Relief', unpublished manuscript in author's possession.

17 See T. Allen and D. Styan, 'A Right to Interfere? Bernard Kouchner and the New Humanitarianism', *Journal of International Development*, 12 (2000): 825–42. While

a heroic achievement at the time, later it was recognized as 'an act of unfortunate and profound folly' in that it prolonged the war for a year and a half and contributed to the deaths of 180,000 people' (I. Smillie, *The Alms Bazaar* (London: IT Publications, 1995); c.f. Allen and Styan, 'A Right to Interfere'.

18 c.f. Allen and Styan, 'A Right to Interfere'.

19 R. Fox, 'Medical Humanitarianism and Human Rights: Reflections on Doctors Without Borders and Doctors of the World', *Social Science and Medicine*, 41 (1995): 1607–16.

20 By the word internationalism, I intend to denote that MSF is structured as a federation of national sections, a product of the organization's history, rather than a truly transnational organization.

21 See P. Redfield, 'Doctors, Borders, and Life in Crisis', *Cultural Anthropology*, 20 (2006): 20, footnote 12.

22 See Fox, 'Medical Humanitarianism and Human Rights'.

23 See Rony Brauman and his comments on this phenomenon in 'From Philanthropy to Humanitarianism: Remarks and an Interview', *South Atlantic Quarterly*, 103 (2004): 406.

24 Renée Fox writes that the United States branch of MSF pays more attention to cooperation with indigenous medical organizations and to learning about the history and cultural background of the places in which they intervene; the French branch (and the headquarters) have been less inclined to learn about cultural differences because they work with a notion of universality that they believe rises above cultural particularities (Fox, 'Medical Humanitarianism and Human Rights').

25 C. Herlizch, 'Professionals, Intellectuals, Visible Practitioners? The Case of 'Medical Humanitarianism', *Social Science and Medicine*, 41 (1995): 1617–19.

26 Allen and Styan, 'A Right to Interfere?', 828.

27 Guilot, 'France, peacekeeping and humanitarian intervention', *International Peacekeeping*, 1 (1994): 31.

28 See *Ibid.*; Fox, 'Medical Humanitarianism and Human Rights'; Brauman, 'From Philanthropy to Humanitarianism'.

29 This debate is ongoing; I attended the annual 2005 international MSF General Assembly meeting in Paris, and again, this question was raised and discussed.

30 Brauman, 'From Philanthropy to Humanitarianism', 410.

31 Allen and Styan, 'A Right to Interfere'.

32 In an interview in 1999, Kouchner referred to the General Assembly Resolutions 43/131 and 45/100 and stated that 'I was not only influential. I was writing it. They were my people, coming from my cabinet and myself'. Quoted in *Ibid.*, 835.

33 *Ibid.*

34 For more on MSF's current status as a transnational entity, see Redfield, 'Doctors, Borders, and Life in Crisis'.

35 See for instance F. Terry, *Condemned to Repeat? The Paradox of Humanitarian Action* (Ithaca: Cornell University Press, 2002), for what Terry calls the 'paradox of humanitarian action': that humanitarianism can contradict its fundamental purpose by prolonging the suffering it intends to alleviate. See also F. Weissman (ed.) *In the shadow of 'just wars'*; and D. Rieff, *A Bed for the Night: Humanitarianism in Crisis* (New York: Simon & Schuster, 2002).

36 Kouchner discussed the complex relationship between NGOs, the media and policy makers in his book *Charité Business* (Paris: Le Pr Aux Clercs-Belfond, 1986).

37 Fox, 'Medical Humanitarianism and Human Rights'.
38 T. Keenan, 'Mobilizing Shame', *The South Atlantic Quarterly*, 103 (2004): 442.
39 For a particularly insightful discussion of the role of media in producing suffering as a distant spectacle upon which one feels no compunction to act, see L. Boltanski *Distant Suffering: Morality, Media and Politics* (Cambridge: Cambridge University Press, 1999).
40 Interview with Tim Allen, April 1999, c.f. Allen and Styan, 'A Right to Interfere'.
41 See K. Blancet and B. Martin (eds) *Critique de la raison humanitaire* (Paris: Le cavalier bleu, 2005) and in particular, the essay by Egbert Sondorp, comparing the different ethical approaches taken by French and 'Anglo-Saxon' humanitarian organizations.
42 Redfield, 'Doctors, Borders, and Life in Crisis', 343.
43 For a detailed discussion of the place of témoignage, see P. Redfield, 'A Less Modest Witness: Collective Advocacy and Motivated Truth in a Medical Humanitarian Movement', *American Ethnologist*, 33 (2006): 3–26.
44 For an interesting juxtaposition of French and British humanitarian traditions, see R. Brauman 'Preface' in Blanchet and Martin (eds) *Critique de la raison humanitaire.*
45 B. Granjon, 'Mission France esiste encore' (Editorial), *Les Nouvelles*, 32(2) (1993); c.f. Fox, 'Medical Humanitarianism and Human Rights', 1614.
46 See B. Taithe, 'Reinventing (French) Universalism: religion, humanitarianism and the "French doctors"', *Modern & Contemporary France*, 12 (2004): 147–58; A. Finkelkraut, *L'humanité perdue: essai sur le XXIe siècle* (Paris: Seuil, 1996).
47 Redfield, 'Doctors, Borders, and Life in Crisis', 337.
48 For a discussion of how an instrumentalist approach to understanding French imperialism – one that assumes civilizing the natives was a cover for baser motives of greed and power – does not take seriously enough the civilizing language which justified intervention, and the truly universal ideals on which it was based, see A. Conklin 'Colonialism and Human Rights, A Contradiction in Terms? The Case of France and West Africa, 1895–1914', *American Historical Review*, 103 (1998): 419–42.
49 Of course, most subjects could never be quite civilized *enough* – Fanny Colonna demonstrates the tension at the heart of French imperial project that kept colonial subjects at an 'appropriate' distance – the best subjects were not too distant from French cultural norms, yet nor were they too familiar with them. See F. Colonna, 'Educating Conformity in French Colonial Algeria' in F. Cooper and A. Stoler (eds) *Tensions of Empire* (Berkeley: University of California Press, 1997). Similarly, Ann Stoler argues that domination consisted in both distancing and incorporation. See A. Stoler, 'Sexual Affronts and Racial Frontiers: European Identities and the Cultural Politics of Exclusion in Colonial Southeast Asia' in Cooper and Stoler (eds) *Tensions of Empire.* For other excellent scholarship on colonialism and its exclusions see for example, E. Said, *Orientalism* (New York: Vintage, 1978); N. Dirks, 'The Policing of Tradition: Colonialism and Anthropology in Southern India', *Comparative Studies in Society and History*, 39 (1997): 189–212; G. Spivak, 'Can the Subaltern Speak?' in C. Nelson and L. Grossberg (eds) *Marxism and the Interpretation of Culture* (Urbana: University of Illinois Press, 1988); F. Fanon, *The Wretched of the Earth* (New York: Grove Press, 1963).
50 L. Dubois 'La République Métissée: Citizenship, Colonialism, and The Borders of French History', *Cultural Studies*, 14 (2000): 26.
51 *Ibid.*, 27

52 Ellis, quoted from Elbourne 1991 'To Colonize the Mind': Evangelical Missionaries in Britain and the Eastern Cape, 1790–1837, DPhil Dissertation, Oxford University, p. 311. c.f. Lester, 'Obtaining the 'due observance of justice'. See also Conklin, 'Colonialism and Human Rights'.
53 For a discussion of how humanitarian and peace-keeping missions feed into nationalist projects, see S. Razack's *Dark Threats and White Knights: The Somalia Affair, Peacekeeping and the New Imperialism* (Toronto: University of Toronto Press, 2004).
54 Redfield, 'Doctors, Borders, and Life in Crisis', 350.
55 Fassin, 'Une double peine'.
56 *Ibid.*
57 K. McClure, 'Taking Liberties in Foucault's Triangle: Sovereignty, Discipline, Governmentality and the Subject of Rights', in A. Sarat and T.R. Kearns (eds) *Identities, Politics, Rights* (Ann Arbor: University of Michigan Press, 1995).
58 U. Mehta, 'Liberal Strategies of Exclusion', *Politics & Society*, 18 (1990): 427–54; U. Mehta, *Liberalism and Empire: A Study in Nineteenth-Century British Liberal Thought* (Chicago: University of Chicago Press, 1999).
59 Redfield, 'Doctors, Borders, and Life in Crisis', 346.

8

Screening out Diseased Bodies: Immigration, Mandatory HIV Testing, and the Making of a Healthy Canada

Renisa Mawani

From the late nineteenth century onwards, health has been a technology of governance constitutive of national borders and racial boundaries. As many scholars have documented in various geographical contexts, nineteenth and twentieth-century public health policies have been intricately linked to racialized nation-formation in several ways. Whereas disease and ill-health were often the racial mark of the 'colonized' and 'uncivilized', the racialized concept of (European) citizenship was historically imagined through ideas around health and vitality.[1] Today, as we move into the twenty-first century, public health remains an imperative of nation-formation. If contagion was historically seen as 'the dark side of the civilizing mission' as Michael Hardt and Antonio Negri claim, in the twenty-first century contagion remains a constant and present danger, but is now the dark side of globalization.[2] Global flows of knowledge, capital, migrant labor, and travel – and the rapid speed at which these now occur – have opened up even greater possibilities for the transmission of germs and disease. 'If we break down global boundaries and open up universal contact in our global village', ask Hardt and Negri 'how will we prevent the spread of disease and corruption?'[3]

This chapter explores the recently enacted mandatory HIV/AIDS testing initiative for all prospective immigrants seeking entry into Canada, a problematic policy that has received surprisingly little critical scholarly attention.[4] Although health screening was expanded to include HIV/AIDS in 2002, the Canadian government's decision to test all immigrants for HIV/AIDS is not a new development. In 1994, the then Reform Party's Immigration critic introduced a motion to Federal Parliament calling for mandatory HIV testing for all persons applying for immigrant status in Canada.[5] During this time, Citizenship and Immigration Canada was already planning a considerable restructuring of immigration law. As part of this process, they consulted extensively with Health Canada about the need for more rigorous health screening procedures. Although Health Canada initially advised that testing all prospective immigrants for HIV/AIDS and excluding those who test positive would be the safest and most productive public health strategy,[6] in light of criticism from various immigrant/refugee and HIV/AIDS advocacy groups, they later revised their recommendation. In April 2001, Canada's Minister of Health sent a newly drafted statement to Elinor Caplan, then Minister of Citizenship and Immigration. The amended opinion was as follows: 'mandatory testing for HIV is necessary, but prospective immigrants with HIV, after receiving counseling need not be excluded from immigrating to Canada on public health grounds'.[7]

On 15 January 2002, the federal government included mandatory testing for HIV/AIDS as part of the new *Immigration and Refugee Protection Act*.[8] Medical testing for a variety of diseases including syphilis, tuberculosis, and more recently, hepatitis B and HIV/AIDS is now a requirement that must be met by almost all prospective immigrants.[9] Although persons who test positive cannot lawfully be denied entry on the basis of public health grounds alone, prospective immigrants may be turned away if seen to pose a 'public health risk' or 'excessive demand' on Canada's already overburdened health care and/or social services network. Interestingly, no other categories of entrant such as tourists, visitors, or returning citizens are required to undergo compulsory HIV/AIDS screening unless they meet specific criteria. When asked about the selective testing procedures Elinor Caplan explained that HIV/AIDS testing would not be feasible for the millions of visitors and returning citizens and residents who enter the country each year. Because of globalization and access to travel, Caplan explained, '[w]e know that it is impossible to shrink wrap our borders'.[10]

Many social theorists have argued that we have now entered into a post-national historical juncture in which the nation-state is increasingly becoming a geopolitical formation of the past. Zygmunt Bauman among others has contended that in a global world, the nation-state has 'lost much of its past allure as a location for secure and profitable investment'. Others have insisted that globalization has stripped the nation-state of many of its former powers, and in the process has placed national formations and sovereignty into question altogether.[11] Etienne Balibar observes that: 'We are henceforth in a situation in which the question of knowing what the terms "nation", "national", and "nationalism" mean, and the idea of inscribing the relation between the individual and the national model have become distinctly more obscure'.[12] However, Balibar and others caution that globalization is not a new phenomenon.[13] Rather, he observes that the nation-form has never stopped transforming itself. A given nation, Balibar explains is 'certainly no longer a "nation" in the same sense of the word as it was two hundred years or even two generations ago'.[14] While the nation-state has undeniably undergone significant shifts, what arguments about globalization and the erosion of the nation-state obscure is the multiple ways in which nations have reconfigured their national borders, often times in direct response to anxieties of a borderless world. While crime and terrorism have (re)emerged as strategies of border control – fears that have been used explicitly to regulate immigration in racialized terms[15] – public health has also resurfaced as an issue of national security and once again figures prominently in reinscribing national borders by determining who belongs inside and outside the nation.

Fears of HIV/AIDS have become particularly salient in discussions of globalization and border control. In her most recent book, *Globalizing AIDS*, Cindy Patton argues that AIDS has had a significant impact upon the ways in which we understand globalization. With the emergence of AIDS, she explains, 'global proximity no longer promised wondrous cultural explorations; rather, it seemed to facilitate the spread of exotic new diseases that were not only deadly to individual bodies, but also threatening to the body politic'.[16] As HIV/AIDS enters into its third decade, the emphasis on the global nature of the disease has become increasingly apparent.[17] Importantly, the construction of HIV/AIDS as a global phenomenon has been partially underpinned by assumptions about the dangers of mobility. In many Western nations, HIV/AIDS with its multiple racialized and sexualized meanings has increasingly come to represent a foreign threat imported into the nation from without. In Canada, the emphasis on compulsorily testing prospective immigrants and not visitors and returning citizens reinforces this discursive construction of foreign-ness with serious material implications. Specifically, the prophylactic response to HIV/AIDS is increasingly and problematically conceptualized in terms of national security as opposed to global access to adequate health care, safe-sex practices at home and abroad, and preventative education.

In this chapter, I situate Canada's new mandatory HIV/AIDS testing policy in the context of globalization and national sovereignty. I argue that compulsory HIV/AIDS testing and health screening more generally have enabled the Canadian nation to reconstitute itself and assert its sovereignty in a globalizing age, when national borders have become increasingly porous. In the first part of the chapter, I briefly sketch out some of the debates about globalization and the erosion of national sovereignty. While I agree that the nation-state has transformed itself in relation to transnational flows and forces, I argue that the border is still an important site through which nations assert control over racialized populations, albeit in different ways.

In the second part, I discuss Canada's new *Immigration and Refugee Protection Act* and the medical inadmissibility provisions in more detail. Here, I unpack the neo-liberal discourses through which health exclusions have been constituted and justified. I suggest that recent discussions about health screening are premised on an explicit 'racelessness'[18] that reflects the assertion of a new Canadian national identity – a 'generous' and 'tolerant' nation; which welcomes all immigrants, white and non-white, but has imposed mandatory health screening for diseases including HIV/AIDS to justifiably protect its limited national resources, including universal health care. In the final section, I problematize the 'racelessness' of Canada's new health screening policies. Notwithstanding the promises of politicians who argue that immigration and diversity are integral parts of Canadian nationalism, I argue that race and racism continue to underpin Canada's immigration law, including health screening. I conclude the chapter with a brief discussion of how racisms and global racial inequalities infuse public health strategies at the border.

Reinscribing borders/reconstituting nation in an age of global flows

Globalization – albeit a contested term – has raised important questions about the role of the nation-state and its capacity to assert and maintain its sovereignty. Transnational flows of information, goods, communication, travel, migration, and trade regimes, have prompted many scholars to ask difficult questions about whether the nation-state can now exist solely with national interests in mind.[19] Zygmunt Bauman has argued that the world has become interconnected and inter-dependent as never before. Bauman observes that, '[f]ew if any nation-states can be said ...to be autonomous, let alone self-sustained and self-sufficient – economically, militarily and culturally'.[20] Similarly, Hardt and Negri explain that transnational flows have dramatically altered international configurations of power. Alongside global markets and global production, they argue, we have seen the emergence of a new global order that has altogether eroded the nation-state and has thus transformed national sovereignty. Hardt and Negri have contentiously called this 'new global form of sovereignty', Empire.[21]

Other scholars have criticized arguments about the newness of globalization and the emergence of a recent global order and instead have taken a much more cautious approach. Masao Miyoshi has critiqued the widespread and ambiguous use of the term 'globalization'. 'If globalization means that the world economy is a seamless unity in which everyone equally participates in the economy', he explains, 'obviously globalization has not taken place'. He adds that 'if globalization means merely that parts of the world are interconnected, then there is nothing new about this so-called globalization: it began centuries ago, as Columbus sailed across the Atlantic, if not earlier'.[22] Others have more explicitly rejected arguments that the nation-state has eroded, pointing out that despite its permeability, the border has a real materiality and remains a technology of in/exclusion aimed primarily at curtailing or preventing the mass migration of peoples.

In *Globalization and its Discontents*, Saskia Sassen contends that national sovereignty is still clearly evident in the governance of immigration.[23] The border and the individual 'as the sites for regulatory enforcement' she explains, remain central to national immigration policy despite the influences of a global economy. Benita Parry has also emphasized the centrality of the border, observing that, 'where inequalities persist, so do borders remain in place and so are flows of populations, cultures, and socialities distorted'.[24] Like Parry, Balibar rejects the idea that the border is losing its significance. Instead, he suggests that the border is 'undergoing a profound change in meaning', and that the 'borders of new sociopolitical entities...are no longer situated at the outer limits of territories'. Rather, he suggests that the border is 'dispersed a little everywhere, wherever the movement of information, people, and things is happening and is controlled – for example in cosmopolitan cities'.[25]

Despite transnational flows of various types, the border does indeed remain a critical site of inclusion and exclusion. Balibar's discussion of the border as a slippery term yet one 'rich in significations',[26] is especially provocative and useful in the context of Canadian immigration and health testing policy, where the state continues to assert its sovereignty through direct control over geographical boundaries. It is important to note however, that health is not a new technology of rule per se. On the contrary, when we look historically, we see that health has always been central to the making of the nation-state, determining who can come in and who must stay out. Writing about Australia, Alison Bashford observes that nation formation 'has found one of its primary languages in biomedical discourse, partly because of the political philosophy which thinks of the population as one body, the social body, or the body of the polity'.[27] Similarly, Alan Petersen and Deborah Lupton have argued that health has been and remains central to citizenship, to one's inclusion and place in the nation. As Petersen and Lupton explain: 'Good health is required for a person to become a "good citizen", for ill health removes individuals from the work force and other responsibilities and places an economic burden on others'.[28]

We see these linkages between borders, citizenship, and foreignness clearly in the history of immigration and health regulation in Canada. Health was historically central to Canadian border control and national security, albeit in different ways. For instance, Canada's first *Immigration Act*, passed in 1869, set out the legal parameters for excluding those deemed physically, morally, and mentally 'unfit' for citizenship.[29] Immigration exclusions based on health grounds were racialized from early on, often explicitly targeting people of color, most notably Chinese laborers, who were routinely forced to undergo medical testing at the border, and were sometimes placed in detention or quarantine confinement.[30] Later in the twentieth century however, health technologies – albeit still racialized – shifted to a process that Nayan Shah has referred to in the US context as 'screening for the fitness of future citizens'.[31] During this period, the state was increasingly concerned with ensuring that *all* immigrants who gained entry into the nation were physically and mentally fit to perform and fulfill their civic duties.

While immigration officials did indeed screen out diseased bodies at the border historically, what we see today is that the border has been transformed and shifted, just as Balibar claims. As I discuss below, health has once again become an important and uncontentious imperative of nation-formation.[32] Under the current health testing and medical screening regime, prospective immigrants must be screened and tested for HIV/AIDS and other diseases *before* they reach the Canadian border. While Canada's early immigration system appointed immigration officers at ports of entry to determine who could come into the country, from the late twentieth century onward, under more recent immigration laws, the border has been dispersed beyond conventional territorial boundaries. Local doctors selected and approved by Citizenship and Immigration Canada – have now become responsible for conducting medical examinations including HIV/AIDS testing when prospective immigrants first apply for entry into Canada. The Canadian border has thus extended beyond ports of entry, placing medical surveillance as far back as the 'front line', in the countries of origin from which prospective immigrants apply for entry.

Health testing, as I have suggested above, is not a new phenomenon. However, there are specific global and national developments, which may lead us to conclude that this process has changed over time. Saskia Sassen's work is particularly useful.[33] She argues that national efforts to regulate immigration have enabled nation-states to assert political and territorial authority over their borders during global times. Although policing immigration through the border has long been a strategy of nation-formation in Canada and the US, Sassen observes that today's border control strategies are 'radically different from past periods'.[34] She identifies three main features that characterize these new conditions. The first is the effect of globalization; nation-states are facing economic conditions that have in many ways neutralized the border and have transformed state sovereignty. Second, the emerging global concern with international human rights has meant that the state is no longer 'the exclusive subject of international law'.[35] And finally, nation-states under the rule of law confront a number of juridical and constitutional changes that have strengthened the rights of citizens and the role of civil society.[36] Each of these factors, Sassen argues, has limited *how* the nation-state can police its borders through immigration.

Although Sassen's observations center on the European context, her arguments are also relevant for Canadian immigration law. From the mid-twentieth century onward, the explicit racial assertion of Canadian national boundaries and the complicity of immigration law became increasingly problematic for a number of reasons. In 1971, the Canadian government – under the leadership of then Prime Minister Pierre Trudeau – announced its official policy of multiculturalism.[37] This policy, which was initiated in response to mass migration and to French and Indigenous resistance, represented a newly imagined Canadian nation; premised on the ideals of pluralism and co-existence, this identity was a distinct departure from the 'white settler' society that had been articulated in earlier historical moments.[38] Rather, this new nationalism was intended to enhance and display the diversity of the body politic while simultaneously reinscribing the mythical identity of Canada as an 'immigrant nation'.[39]

Relatedly, border control and the movement of peoples has been internationalized as never before. From the late twentieth century onwards, liberal democratic states including Canada, aspiring to be players on the world stage, have no longer been able to justifiably restrict immigration in racial, or more accurately, racist terms. Not only would an explicit racism be inconsistent with Canada's national project of multiculturalism, but it would also jeopardize the nation's international reputation as a leader in the field of human rights. This does not mean that race is no longer implicated in immigration and national formation, but rather, that discourses of race have become increasingly ambiguous and difficult to identify. Writing about changes to Canadian immigration law during the 1990s, Sherene Razack explains that: 'In liberal democracies overtly racist acts cannot be tolerated'. However, she contends, 'if the story of an overtly racist act is transformed into the story of a state forced to defend itself from bodies bent on betraying its trust, then such acts become acceptable and even laudable'.[40] In Canada, more recent restrictive border regulations have been articulated through a new multicultural, cosmopolitan, and humanitarian language of inclusivity that has masked the explicit and implicit racism that has underpinned the implementation of immigration and health screening policies. Importantly, as I discuss below, these practices have been legitimized as a necessary means of protecting the nation from the dark side of globalization, including disease, crime, and terrorism.

Despite Canada's efforts to enact a 'raceless' immigration regime, it is important to note that calls for more extensive medical admissibility provisions were at least in part influenced by racialized fears of a borderless world. Although Citizenship and Immigration Canada was long planning to restructure Canada's immigration laws, including the health screening procedures, global fears of contagion did buttress demands for more extensive health testing. Two examples illustrate this point. In the summer of 1999, four boats carrying close to 600 illegal migrants from the Fujian province of China arrived off the coast of Vancouver Island. The subsequent news coverage of these arrivals precipitated a 'crisis' in Canada: the nation was thought to be under siege by yet another wave of Chinese migration, this time illegal. While much of the news coverage focused on the migrants as criminals, attention was also paid to the perceived health and national security risks that the migrants posed to Canadian citizens.[41]

Two months after the migrants arrived, the Reform party began making allegations that these 'illegals' posed a potential 'AIDS risk' and that Canada needed to tighten its immigration policies. As one article published in the *Victoria Times-Colonist* cautioned: 'Canadians should be at risk from terrorists and communicable diseases such as AIDS if action isn't taken soon to toughen Canada's immigration law'.[42]

Other media spectacles of foreign bodies bringing exotic germs into the Canadian nation also underpinned racialized debates about how best to reform Canada's immigration policy. In 2001, Colette Matshimoseka, a Congolese woman, arrived at Pearson International Airport suffering from an unidentifiable illness that authorities speculated to be Ebola. Although Matshimoseka was not an immigrant but a visitor to Canada, fears of her black African body infected with Ebola (which was later confirmed a misdiagnosis) did shadow parliamentary debates about Canada's new immigration law, and more particularly, the need for more stringent health screening of prospective immigrants.[43] In fact, fears of diseased and foreign bodies bringing HIV/AIDS and Ebola into Canada were raised by at least one Member of Parliament as a justification for tighter and more stringent border regulations. In governmental discussions about Canada's newly proposed immigration law, John Bryden, a member of the Liberal Party, expressed concerns about health testing exemptions for those prospective immigrants who were applying for entry under the 'family class'. Currently, spouses, common-law partners, and children fall into the 'family class' provisions and are exempt from health-based exclusions.[44] Bryden took issue with the fact that common law partners were not required to undergo health testing under the new law, cautioning that partners with 'a disease like AIDS, or Ebola virus, or any of these things' would still be allowed to enter Canada.[45] That Bryden would name AIDS and Ebola as potentially threatening to the body politic is significant, suggesting that these crises of diseased racialized bodies did in some way impact the psyche of the nation.

While it could easily be argued that Canada's new health testing policies are intended to govern the flow of racialized bodies across the border, the federal government has tried hard to show that these new enactments are really aimed at protecting Canada's over-burdened health care and social services system from 'unhealthy' immigrants who do not deserve the nation's generosity. Many politicians have insisted that these new provisions are fair and non-discriminatory. While Canada's new *Immigration and Refugee Protection Act* does closely monitor the border, screening out those prospective immigrants and refugees thought to be 'diseased' and 'criminal', Canadian politicians have argued that this law is distinct from prior immigration regimes as it recognizes the importance of international migration, particularly the economic and social contributions of prospective immigrants.[46] As several Members of Parliament explained further, the new law could not possibly be exclusionary as it upholds the standards of the *Charter of Rights and Freedoms* and international law.

In her opening address to the Standing Committee, Elinor Caplan promoted the *Immigration and Refugee Protection Act* as follows:

> This is important legislation ... which will be of great benefit to the country. The reason why is quite simple. By saying 'no' more quickly to those who would abuse our rules, we will be able to say 'yes' more often to those immigrants and refugees who Canada will need to grow and prosper in the years ahead.[47]

Caplan added that while the proposed Act made clear that Canada would 'stand on guard', as this 'is not a country that offers safe haven to anyone other than those who need our protection', that the legislation was premised on 'the principles of equality and freedom from discrimination'.[48] Thus, the basis for inclusion/exclusion from Canada would not be contingent upon race and/or national origin as was the case historically, but would now carefully distinguish between those 'deserving' and 'undeserving' immigrants – between those morally and physically fit and unfit for citizenship. However, as some scholars have warned, and as history tells us, narratives of 'good', 'bad', 'fit', and 'unfit' immigrants are in fact deeply racialized.[49]

A new Canada: restricting immigration through 'excessive demand'

Citizenship and Immigration Canada and Health Canada had long been planning a major restructuring of Canada's immigration laws, which many described as outdated and in need of replacement. Since it was first passed in 1976, the *Immigration Act* was amended over 30 times but never entirely revised. Bill C-11, now known as the *Immigration and Refugee Protection Act*, was the first comprehensive piece of legislation aimed at replacing Canada's old immigration law. After more than a year of debate and discussion among Canadians and politicians, the Bill was finally passed by the federal government on 1 November 2001 and became law on 28 June 2002.

Despite the fact that the *Immigration and Refugee Protection Act* is deliberately restrictive and aimed at protecting the Canadian nation through tougher penalties for 'criminals' and others who pose 'security risks', politicians who engaged in discussions about the Act continuously articulated the open and inclusive nature of this new law. Members of the Standing Committee made numerous efforts to explain and defend how this law was distinct from Canada's past efforts to restrict immigration in discriminatory terms. Rather, this new legislation was deemed to value immigration by recognizing that immigrants and refugees 'built this country' and would continue to do so in the future.[50]

Throughout the proceedings, politicians – even members of more conservative political parties – newly constructed Canada as a tolerant and generous nation of immigrants. As the Chair of the Standing Committee, Joe Fontana, explained to the House of Commons, 'immigration has been an absolutely positive asset in helping build this country over the past 130 years or so...Canada has a proud history and tradition of compassion for those bona fide refugees who have been persecuted in their own lands'.[51] Similarly, Stockwell Day echoed these sentiments explaining that the Canadian Alliance – Canada's right-leaning political party – was 'proud of Canada's heritage as a country that welcomes immigrants from all parts of the globe, from all races, [and] from all religions'.[52] These declarations of a new and improved Canada have helped to obscure the ways in which race continues to shadow immigration policies, including restrictive health policies.

In his book, *The Racial State*, David Goldberg reminds us that liberal democracies often deny the ways in which the 'hidden hand of the state' promotes, maintains, and manages the racial order.[53] While these processes remain pervasive, Goldberg explains that they have now 'recede[d] into the background' becoming less formal:

> The racial state gave way to the raceless one, the race-bound one to the supposedly race-blind one. But it could do so precisely because racial conditions and presuppositions penetrated [the] social order so extensively, becoming routinized administratively and in everyday life. Racelessness, in short, traded on the fact that race became so readily, one might say universally, assumed.[54]

In the realm of Canadian immigration law, the mythical immigrant nation that values the contributions of 'good' immigrants and excludes 'bad' immigrants trades on a particular currency of race. As Sherene Razack reminds us, '[p]eople who cross borders are increasingly divided into high- and low-risk groups, with an effort made to facilitate the crossing of the low-risk while devoting extra resources to the policing of the high-risk group'.[55] Although these risks are not explicitly or overtly about race, we know for example, that 'high risk' populations in the history of HIV/AIDS discourse have been racialized, gendered, and sexualized.[56]

It is precisely this logic that is at work in Canada's new immigration regime. By making clear distinctions between the 'good' and the 'bad', the 'deserving' and the 'undeserving', the Canadian government has justified the need for tighter policies while simultaneously upholding Canada's national narratives of freedom and equality. In other words, this logic enables Canada to be restrictive on the one hand – by excluding those who may be suspected of crime, terrorism, and disease – while remaining open on the other. Elinor Caplan explained to one Liberal MP that Canada is competing in a global economy for the best and brightest immigrants. 'There are other countries now that are facilitating and trying to entice highly skilled immigrants, because they recognize the economic importance of human capital and smart people'. However, she continued that 'Canada is known around the world as a country that is open. We have an open door and a transparent selection point system. We have...shamelessly attracted the best and the brightest, the cream of the crop, from around the world, and we will continue to do that'.[57] But, including the 'best and brightest' means making distinctions between those immigrants who will enrich the nation and those who will unsettle it.

Along with various other exclusions in the *Immigration and Refugee Protection Act*, Canadian authorities welcomed the health screening provisions of Bill C-11 as a necessary and much improved policy change. Although the proposed amendments were intended to update Canada's medical surveillance program by adding mandatory tests for a number of diseases including HIV, Elinor Caplan assured the House of Commons that these changes were not intended to be discriminatory. As Caplan explained the 'point is to modernize our routine testing system, not to make it disease specific'.[58] In a later hearing, Joan Atkinson, then Assistant Deputy Minister for Policy, elaborated that 'modernizing' meant 'determining whether there are additional routine tests we should be putting into place, which will include HIV, and others too, such as Hepatitis B, such as Chagas' disease – potential public health issues in those *areas of the world where we select our immigrants and refugees*'.[59] Thus, the objectives of the proposed mandatory HIV/AIDS testing program was not intended to restrict entry to individuals who tested positive, but to keep up with the global spread of disease and with international health policy changes; even though many countries that have mandated HIV/AIDS testing for prospective immigrants do carefully police their borders in racially discriminatory ways.[60]

Under the new law, the categories of inadmissibility are conveniently consolidated into one section, section 38. There are two notable changes here. While Citizenship and Immigration Canada added HIV and Hepatitis B to its list of diseases for which testing must occur, they eliminated the initial argument that immigrants who test HIV positive constitute a 'public health risk' and should be excluded. Now, potential immigrants who test positive for HIV/AIDS can only be excluded if they are thought to pose an 'excessive demand' on the health and social services system. Both of these changes were intended to move medical surveillance away from a discretionary system (under the old *Immigration Act*) to a new economic model that left little room for interpretation by over-worked and poorly trained medical personnel. Writing about Europe, Balibar cautions us against accepting these economic and seemingly 'non-discriminatory policies' of border control. He explains that Europe 'has become conscious of the positive value of the other as such, but it keeps excluding people by systematically combining criteria of culture (practically equivalent to race) and economic discrimination'.[61] As I discuss in the following section, culture and economy have become the new discourses of exclusion for those who test HIV positive.

Many politicians have agreed that the new medical inadmissibility provisions and health screening policies under the *Immigration and Refugee Protection Act* have significantly improved. While the new law purports to reduce the discretion of medical officers, it also sets out more specific and explicit criteria for exclusion. Under Canada's previous immigration law, while testing for HIV/AIDS was not mandatory, medical doctors examining prospective immigrants could request an applicant be tested for any illness. Under this system, as Elinor Caplan explained in the Standing Committee hearings, doctors had 'complete discretion...to order any test to determine whether a person is well enough to enter Canada and whether they are medically admissible to Canada'.[62] Some critics who made submissions to the Standing Committee raised concerns about this unrestrained power given to overseas immigration officers. Michael Battista, a Toronto-based immigration lawyer working for the organization, Equality for Gays And Lesbians Everywhere (EGALE) Canada, was among those who argued that the sweeping powers of immigration officers opened up a number of problems, including racial profiling. 'The testing of Africans for immigration to Canada', Battista explained, 'is highly discretionary. It's not dependent on anything that's really in the act. It's a decision a doctor takes in examining an applicant to Canada for immigration'. Battista added that, 'a lot of these doctors don't have adequate counseling or training to deal with HIV. I've had applicants who have been told over the phone that they're HIV-positive. It's had a devastating effect on relationships'.[63]

Recognizing these problems, Canada's new immigration law promised to move away from a discretionary system to an economic model based on 'objectivity'. To begin with, HIV/AIDS testing would now be mandatory for all prospective immigrants, thereby eliminating the problem of racial profiling. Since all prospective immigrants would now be required to take a test, medical health officers would no longer be able to impose their own judgments about which bodies and regions of the world were most risky. Second, the new law would set out explicit definitions of excessive demand. Because excessive demand was not defined in the previous law, but was once again contingent upon the interpretation of medical and immigration officers, critics argued that these criteria could not be interpreted 'objectively'. While some questioned whether doctors working overseas were knowledgeable enough to make these decisions,[64] others argued that these provisions were extremely narrow, focusing only on what immigrants would take from a generous nation, rather than what they would give. When asked by one Member of Parliament as to what constituted excessive demand, Joan Atkinson, explained that Bill C-11 would be 'moving from the current system, which is based very much on information and the knowledge of the medical officer, to an objective, cost-based model'.[65]

Exclusions based on excessive demand were added to the old immigration regime in 1976. These provisions were intended to protect Canada's universal health care system, which had been introduced less than a decade earlier. Under the old *Immigration Act*, prospective immigrants could be deemed medically inadmissible if thought to be a threat to public health and/or if they posed an excessive demand on Canada's health care or social services system. One of the main problems with the old law was that it did not include a definition of excessive demand. Many critics argued that doctors working for Citizenship and Immigration Canada had little expertise to determine whether an applicant would in fact overburden the health and/or social services programs.[66] Under the *Immigration and Refugee Protection Act*, excessive demand was to be determined on the basis of a 5-year window, 'where a person has a condition or a disease that is likely to cause demands higher than the average annual cost for Canadians on the health care system'.[67] Statistics Canada estimated that the average person uses approximately $2,800 in health care services per year. A 5-year threshold would thus be $14,000. Consequently, anyone who may cost the system more than that could potentially be denied entry.

Although many observers were satisfied with the new provisions, Alana Klein from the HIV/AIDS Legal Network expressed several concerns about the definition of excessive demand. 'First of all', she offered, 'we're very pleased to see that finally there's a definition of the term "excessive demand on health care and social services".' However, Klein explained that the provisions suggest that for chronic illnesses, including HIV, the window of projected costs could extend to 10 years. A 'ten year projection period is inappropriately long, especially in the case of HIV/AIDS' as the 'costs for treatment are extremely variable over time'.[68] Despite the Standing Committee's assurances that the new Act would be less discretionary than its predecessor, Joan Atkinson explained that in determining excessive demand, every case 'is examined on its own circumstances, and an evaluation is made on what the particular prognosis is, the nature of the disease, the status of the individual, and the treatment of the individual will need to receive in Canada'.[69] In other words, despite discussions about objectivity and neutrality, doctors and immigration officers would still be making case-by-case determinations on who could enter the nation.

Throughout discussions about Bill C-11 including its health screening policies, Canadian politicians – and to some extent advocacy groups and organizations – congratulated the Standing Committee for producing a proposed law that balanced the safety and protection of Canadians with the human rights of immigrants and refugees. As I have suggested above, the exclusionary provisions of the new Act were continually rendered non-discriminatory and 'raceless', reflecting a distinct departure from previous immigration regimes on the one hand, while serving as evidence of a new global, multicultural, and cosmopolitan Canada on the other. However, several scholars have argued that seemingly neutral concepts often draw upon historical and contemporary racial inequalities that structure both the global and the local. Writing about disease, Sander Gilman reminds us that: 'Like any complex text, the signs of illness are read within the conventions of an interpretive community that comprehends them in the light of earlier powerful readings of what are understood to be similar or parallel texts'.[70] Although the language of exclusion has been neutralized, excluding potential immigrants from entry into Canada remains contingent upon the nation's prior (and ongoing histories) of colonialism, racism, and xenophobia.

Raceless states as racial states[71]

The exclusion of prospective immigrants who test positive for HIV/AIDS has largely been articulated through a cosmopolitan language of inclusion on the one hand and protection of Canada's universal health care system, on the other. Thus, as I discuss above, Canadian politicians have been careful to frame medical inadmissibility as an economic issue as opposed to a racial or homophobic one. Jay Hill, a Canadian Alliance Member of Parliament for Prince George-Peace River echoed these sentiments in his weekly editorial:

> As Canadians, we should have the right to determine who can seek entry into our country. There is nothing wrong with linking immigration to economic factors, such as employment levels, job skills or language ability. The government must also consider the economic repercussions of admitting people we know may become a burden on our health care system down the road.[72]

As Hill notes, the federal government – regardless of political orientation and ideology – did indeed frame immigration, and Canada's need for immigrants, as an economic issue. If inclusion was to be determined economically, Hill and others explained, then exclusion should also be contingent upon whether prospective immigrants may potentially drain the public purse. Inky Mark, the Canadian Alliance immigration critic emphasized this point, insisting that Canada's decision to implement mandatory testing was purely an economic one. He explained that in 2000 alone, approximately 200 HIV-positive persons entered into Canada and had already cost the federal government $26 million.[73]

Notwithstanding the inclusive and neutral language of Canada's *Immigration and Refugee Protection Act*, including the universal nature of health based exclusions, some observers have pointed out that these provisions remain discriminatory. Specifically, the economic discourse of immigration and health care that has been used to justify mandatory testing for various diseases becomes particularly problematic in the context of HIV/AIDS. When we talk about excluding those who test positive – even if on economic grounds – we are in effect talking about restricting entry to racialized (black) bodies that come from specific regions of the world, most notably the African continent.[74] In the health exclusions of immigration policy, race and racism are not explicitly present but in fact work through their invisibility in various ways. Racism is articulated through 'coded signifiers' ('disease' and 'economy'), discursive practices that David Goldberg so long ago described as being 'a central feature of the concept of race in modernity'. He argues that neutral signifiers hide their discriminatory or exclusionary dimensions behind a 'universal characterization',[75] although the outcome or effects of these may be clearly racialized.

To begin with, prospective immigrants are currently the only group in Canada required to undergo mandatory HIV/AIDS testing. As the Canadian Council of Refugees has pointed out, this policy is reminiscent of past immigration initiatives in which foreigners were viewed to be carriers of disease. What makes the new discourse distinct, however, is that Canada can no longer denounce potential immigrants as public health threats, as was the case in earlier historical moments. To do so would not only tarnish Canada's reputation on the world stage, but also would contradict the new national mythology of the multicultural immigrant nation.

Moreover, to see potential immigrants who test HIV positive as posing an excessive demand on the health care system is also a racialized narrative with historical roots. Excessive demand is highly suggestive of an earlier discourse, that of the 'immigrant as welfare cheat', who is not interested in what s/he can contribute to the nation but is only interested in what Canada can give. African American scholars have consistently pointed out that the image of the 'Welfare Queen' personifies all of America's racial problems. Writing of Anita Hill and Clarence Thomas, Wahneema Lubiano explains that, 'urban crime, the public schools, the crack trade, [and] teenage pregnancy are all narratives in which [the] welfare queen is writ large'.[76] In Canadian immigration discourse, anxieties about welfare and welfare fraud are similarly racialized. Commenting on the 'Identity Documents' amendments to the *Immigration Act*, Sherene Razack notes that the 'welfare-fraud immigrant, the criminal immigrant, and the bogus refugee are all key figures'.[77] Like the 'Welfare Queen', the welfare-fraud immigrant is one who has no past but only a future. Razack explains: 'In this story, a refugee who gets to our borders without identity documents is someone we do not know, and someone who is likely to defraud us, as well as someone who is duped by smugglers'. Importantly, her 'character and history are fixed for us in the moment of the encounter at the border. She is not someone with a past – although she is someone with a guessed-at future – that of welfare abuser'.[78]

Like Lubiano's 'Welfare Queen' and Razack's 'welfare abuser', the prospective immigrant who tests positive for HIV/AIDS is also someone without a past and only a 'guessed-at future'. Colonialism, imperialism, globalization, inequalities between North and South, and lack of access to adequate health care, are all erased from the encounter between the immigrant and the border. Rather, the central question for immigration authorities is whether this person will be a future drain on Canada's health care system: whether she will take advantage of Canada's generosity. Lubiano's analysis is instructive here. She argues that while the welfare queen 'represents moral aberration and an economic drain...the figure's problematic status becomes all the more threatening once responsibility for the destruction of the "American way of life" is attributed to it'.[79] This is precisely the argument that Canadian authorities have made in support of HIV/AIDS testing: that while 'undesirable' immigrants may indeed be morally inept their potential to drain the Canadian economy makes them a threat to the 'Canadian way of life' and to Canada's future. Although race is obscured in these accounts, narratives of excessive demand are indeed racialized as welfare cheats/abusers/those who take advantage are characteristically people of color from the 'Third World'. Moreover, these stories function to justify the vigilant policing of the border against those unhealthy bodies that may eventually become economic parasites on the nation-state.

Throughout their discussions of Bill C-11, the Standing Committee and Members of Parliament have recognized that Canada, and more particularly Canadian immigration policy, exists within a global context. In terms of disease and health screening, Citizenship and Immigration Canada have acknowledged that the transmission of germs and disease, including HIV/AIDS is also a global problem. In response to the global crisis surrounding HIV/AIDS it is important to note that Canada has taken an important role by pledging $120 million over 3 years to fight the disease in Africa and other countries where AIDS is thought to be rampant.[80] Interestingly, the Canadian government has agreed to fight HIV/AIDS 'over there', yet the government's response to prospective immigrants who test HIV positive and who may burden the health care system does not reflect the same standards of humanitarianism. On the contrary, the Canadian government's international and national response to HIV/AIDS simply reinscribes the vast global inequalities between North and South that can be traced back to colonialism. Although colonialism – in the formal sense – is now over, Cindy Patton points out that the global management of health 'sustains poor countries' dependence on European and American superpowers'.[81] Patton asks us to think critically about the concept of World

Health. 'Even in the noble effort to globalize basic human rights', she explains, 'the very concept of "world health" teeters ambiguously between a democratic ideal and a genocidal fantasy: Does it mean a world in which health is distributed, or one from which the unhealthy have been eliminated by any means necessary?'[82]

Canada's response to the world health crisis, and specifically to HIV/AIDS, does indeed teeter ambiguously. While the Canadian government is committed to improve the health of the world's poorest nations, it is simultaneously intent on preventing diseased people from impoverished countries from entering into Canada, and thus threatening the nation's valuable health care resources. Arguably, while these initiatives do appear to be democratic, upon more careful scrutiny, one could argue that these policies are aimed at preventing the 'unhealthy' from gaining access to Western countries, where the standards of health care are significantly better. Furthermore, keeping diseased bodies 'over there' works to reaffirm – both discursively and materially – that HIV/AIDS is a foreign (read Third World) problem whose prevention requires foreign aid and vigilant governance at the borders of Western nations.

In 2003, just one year after the medical admissibility provisions of the *Immigration and Refugee Protection Act* took effect, the federal government denied entry to 75 prospective immigrants who tested positive for HIV/AIDS. However, another 207 potential immigrants who also tested positive were allowed entry into Canada.[83] In response to these practices, some medical practitioners have predicted that these exclusions will have the most damaging effects upon immigrants from regions across Africa. Phillip Berger, a Toronto doctor who has treated AIDS patients for 20 years, told the *Globe and Mail* that the mandatory testing policy will have a 'disproportionately punitive effect on the highest prevalence of HIV, which are in sub-Saharan Africa'. He added 'there are racial underpinnings of this policy – its [sic] discriminatory'.[84] Writing about the construction of African AIDS, Simon Watney argues that, 'the language of metaphor that informs so much African AIDS commentary carries a very specific ideological cargo' of European colonialism.[85] Watney elaborates that, in Western narratives about AIDS, Africans are 'set up as authentic "natives" – unreliable, superstitious, in a word, primitive'.[86] What such narratives fail to consider, however, is how 'AIDS in Africa is a symptom of colonialism and exclusion from access to global/local resources'.[87] Importantly, mandatory health screening for HIV/AIDS does little to protect Canada from contagion and does more to exclude African nations from the global economy.

In October 2004, the *Globe and Mail* printed an article on the front page of its Saturday edition entitled, 'Coming to Canada with Dreams and HIV'. While this article reinforced myths that HIV/AIDS is a foreign disease that is imported from without, it also reinforced the hysteria around the need for tighter border control policies. The article explained that the face of Canada's new HIV crisis has shifted and now includes 'immigrants and refugees from regions where the virus is endemic'. Specifically, those from Africa and the Caribbean are now thought to form the 'fourth largest group' of HIV infected persons in Canada. In Ontario, the article continued, immigrants and refugees from these regions rank second among HIV-positive populations, 'fewer than gay men but more numerous than injection drug users'.[88] Notably, the spread of infection is partly explained through the 'primitive culture' of Africans. These newcomers, it is alleged, 'come from sexual cultures where practices such as female genital mutilation and vaginal cleansing may render mainstream prevention programs ineffective'.[89] Here, HIV/AIDS is not only affirmed in racial terms – as an African disease – but also one that is distinctly different from its Western equivalent. Thus, AIDS in Africa is seen as distinct from that in the West, 'with different modes of transmission having to do with dramatic differences in Western and African sexual practices'.[90]

Throughout the Standing Committee's deliberations and public hearings, few observers raised questions about the racialized aspects of Canada's proposed health screening practices. In the few cases where such issues were raised, Standing Committee members promptly dismissed these concerns as unfounded, exaggerated, and erroneous. The African Canadian Legal Clinic was among the few organizations to express anxieties about the racial undertones and implications of Bill C-11. The Clinic has been actively involved in critiquing Canada's immigration policy – past and present – as one that considers people of African descent to be 'undesirable' immigrants.[91] When Erica Lawson, a Policy and Research Analyst for the Clinic raised issues to this effect during the Standing Committee briefings, Joe Fontana, the Chair of the Committee responded to her as follows:

> I understand that we have challenges, especially as it [sic] relates to racism. I looked for color in this particular bill, and I don't see it there. I'm sorry you feel that this bill is race based. I think we've gone out of our way to ensure that this has the right gender balance...We want to make sure it is not colored in any way, shape, or form.[92]

Although Fontana was willing to consider the gender critiques of Bill C-11, his reaction to Lawson illustrates an explicit rejection of racism: that the Bill could not possibly be construed as racist and that any suggestions of the sort had more to do with those who made the allegations than with the proposed legislation itself. This is hardly surprising given than liberalism's response to racism at best fails to take it seriously and at worst, denies racism altogether.[93]

Conclusions

Throughout deliberations about the *Immigration and Refugee Protection Act*, and more specifically, the health screening policies for HIV/AIDS, discussions of race and racism have been virtually absent yet have shadowed discussions in complicated ways. Although Members of Parliament and others have argued that the law is neutral, fair, and non-discriminatory, I have suggested throughout that this cosmopolitan and raceless language has masked the ways in which race and racism underpin mandatory HIV/AIDS testing. Many scholars have argued that racism continues to flourish precisely because central to its existence is a politics of denial.[94] In other words, although race and racism are not explicitly articulated in discussions of medical inadmissibility, if we read between the lines we see that it has an overwhelming presence in discussions of border control during globalized times.

David Goldberg argues that, 'racelessness and globalization are mutually implicated'. He explains that, 'the lunge to globalized frames of reference promotes the retreat from explicit racial reference; and the pull of racelessness disposes subjects to evacuate local racial terms'.[95] Yet if epidemics are to become major threats to global and national security as many have argued, then we need to spend more time interrogating what sorts of social relations we are governing through our global quest for health. If the 'ideologies encoded in AIDS research have laid a more sublime foundation for selecting groups of people for detention and destruction',[96] as Cindy Patton claims, then the mandatory HIV/AIDS testing policy is sure to have dangerous implications for the Canadian nation, enabling Canada to reconstitute itself in racial terms while offering little protection from the importation of disease. As Hardt and Negri point out, HIV tests will not protect national borders in an age of globalization. 'The boundaries of nation-states...are increasingly permeable by all kinds of flows. Nothing can bring back the hygienic shields of colonial boundaries. The age of globalization is the age of universal contagion'.[97]

Notes

An earlier version of this chapter was presented at the Law and Society Annual Meetings in Pittsburg, 2003. I would like to thank participants for their questions. Thanks are also due to Elizabeth Rondinelli for her invaluable research assistance and to the University of British Columbia for an HSS Large Grant that made this research possible. Finally, I would like to thank Alison Bashford for important conceptual and editorial suggestions.

1 See, W. Anderson, 'Excremental Colonialism: Public Health and the Poetics of Pollution', *Critical Inquiry*, 21 (1995): 640–69; D. Arnold, *Imperial Medicine and Indigenous Societies* (Manchester: Manchester University Press, 1988); A. Bashford, *Imperial Hygiene: a critical history of colonialism, nationalism, and public health* (London: Palgrave, 2004); M.E. Kelm, *Colonizing Bodies: Aboriginal Health and*

Healing in British Columbia, 1900–1950 (Vancouver: University of British Columbia Press, 1998); R. Mawani, ''The Island of the Unclean': Race, Colonialism, and 'Chinese Leprosy' in British Columbia, 1891–1924', *Journal of Law, Social Justice and Global Development*, 1 (2003): 1–21; B. Pati and M. Harrison (eds), *Health, Medicine, and Empire: Perspectives on Colonial India* (Hyderabad: Orient Longman, 2001).

2 M. Hardt and A. Negri, *Empire* (Cambridge: Harvard University Press, 2000), p. 135.

3 *Ibid.*, p. 136.

4 For notable exceptions see C. Murdocca, *Foreign Bodies: Race, Canadian Nationalism and the Trope of Disease* (Ontario Institute for Studies in Education, University of Toronto: Unpublished MA thesis, 2002); A. Klein, *HIV/AIDS and Immigration: Final Report* (Canadian HIV/AIDS Legal Network, 2001).

5 The Reform Party was Canada's right leaning political party, which has now been renamed as the Canadian Alliance.

6 Klein, *HIV/AIDS*, p. 39.

7 Cited in Klein, *HIV/AIDS*, p. i.

8 *Immigration and Refugee Protection Act* (2001) c. 27. Hereinafter *IRPA*.

9 Those from the 'family class' are exempt from testing.

10 A. Thompson, 'No Entry For Immigrants with HIV', *Toronto Star* (21 September 2000).

11 Z. Bauman, *Society Under Siege* (Cambridge: Polity Press, 2002), p. 10.

12 E. Balibar, *We the People of Europe?* (Princeton: Princeton University Press, 2004), p. 12.

13 See F. Jameson, 'Globalization as Philosophical Issue' in F. Jameson and M. Miyoshi (eds), *The Cultures of Globalization* (Durham: Duke University Press, 2003), pp. 54–77.

14 Balibar, *We the People*, p. 22.

15 In Canada see S.H. Razack, '"Simple Logic": Race, the Identity Documents Rule and the Story of a Nation Besieged and Betrayed', *Journal of Law and Social Policy*, 15 (2000): 181–209. See also I. Grewal, 'Transnational America: Race, Gender, and Citizenship After 9/11', *Social Identities*, 9 (2003): 535–61.

16 C. Patton, *Globalizing AIDS* (Minneapolis: University of Minnesota Press, 2002), p. x.

17 See M. Raimondo, '"Corralling the Virus": Migratory Sexualities and the "Spread of AIDS" in the US Media', *Environment and Planning D: Society and Space*, 21 (2003): 389.

18 For a discussion of 'racelessness' see D.T. Goldberg, *The Racial State* (Cambridge: Blackwell, 2002), p. 221. In Canada see C. Backhouse, 'Bias in Canadian Law: A Lopsided Perspective', *Canadian Journal of Women and the Law*, 10 (1998): 170.

19 See A. Appadurai (ed.), *Globalization* (Durham: Duke University Press, 2001); Z. Bauman, *Globalization: The Human Consequences* (Cambridge: Polity Press, 1998); Jameson and Miyoshi (eds) *The Cultures of Globalization*; S. Sassen, *Globalization and its Discontents* (New York: New York Press, 1998).

20 Bauman, *Society*, p. 80.

21 Hardt and Negri, *Empire*, p. xii.

22 M. Miyoshi, '"Globalization", Culture, and the University' in Jameson and Miyoshi (eds) *The Cultures of Globalization*, p. 248.

23 Sassen, *Globalization*, p. 7.

24 B. Parry, *Postcolonial Studies: A Materialist Critique* (London and New York: Routledge, 2004), p. 95.

25 Balibar, *We the People*, p. 7.
26 *Ibid.*, p. 7.
27 Bashford, *Imperial Hygiene*, especially chapters 5 and 6.
28 A. Petersen and D. Lupton, *The New Public Health: Health and Self in the Age of Risk* (London: Sage, 1996), p. 65.
29 *Immigration Act. 1869*. 32, 33 Vic c. 10.
30 See Mawani, '"The Island of the Unclean"'; R. Menzies, 'Race, Reason, and Regulation: British Columbia's Mass Exile of Chinese "Lunatics" aboard the Empress of Russia, 9 February, 1935' in J. McLaren, R. Menzies and D. Chunn (eds), *Regulating Lives: Historical Essays on the State, the Individual, and the Law* (Vancouver: University of British Columbia Press, 2002), pp. 196–230; P. Ward, *White Canada Forever: Popular Attitudes toward Orientals in British Columbia* (Montreal and Kingston: McGill-Queen's University Press, 1990).
31 N. Shah, *Contagious Divides: Epidemics and Race in San Francisco's Chinatown* (Berkeley: University of California Press, 2001), p. 180.
32 In the parliamentary debates about Bill C-11 now known as *IRPA*, several MPs raised concerns about the implicit assumptions that immigrants are criminals or terrorists. Interestingly, few raised objections about the health screening initiatives.
33 S. Sassen, *Guests and Aliens* (New York: New Press, 1999).
34 *Ibid.*, p. xvii.
35 *Ibid.*
36 *Ibid.*
37 On Canadian Multiculturalism see H. Bannerji, *The Dark Side of the Nation* (Toronto: Canadian Scholars Press, 2000).
38 See S.H. Razack, ''When Place Becomes Race' in S.H. Razack (ed.), *Race, Space, and the Law: Unmapping a White Settler Society* (Toronto: Between the Lines, 2002), pp. 1–14.
39 For a critique of the 'Immigrant Nation' mythology see B. Honig, *Democracy and the Foreigner* (Princeton: Princeton University Press, 2001); S.H. Razack, 'Making Canada White: Law and the Policing of Bodies of Color in the 1990s', *Canadian Journal of Law and Society*, 14 (1999): 159–84. See also R. Mawani, '"Cleansing the Conscience of the People": Reading Head Tax Redress through Canadian Multiculturalism', *Canadian Journal of Law and Society*, 19 (2004): 127–51.
40 Razack, 'Simple Logic', p. 187.
41 S.P. Hier and J.L. Greenberg, 'Constructing a Discursive Crisis: Risk, Problematization and Illegal Chinese in Canada', *Ethnic and Racial Studies*, 25 (2002): 490.
42 'Reform Sees AIDS Risk', *Times Colonist* (3 September 1999).
43 C. Murdocca, 'When Ebola Came to Canada: Race and the Making of the Respectable Body', *Atlantis*, 27 (2003): 24–31. See also Murdocca, *Foreign Bodies*.
44 *IRPA*, 38(2).
45 Legislative Committee of the House of Commons, 1st Session 37th Parliament 2001–2002. [Hereinafter Legislative Committee] J. Bryden, Standing Committee on Citizenship and Immigration, 13 March 2001, 10:45.
46 These nationalist discourses are evident in the title of the Standing Committee's Final Report on *IRPA*, 'Building a Nation', March 2002.
47 Legislative Committee, E. Caplan, Standing Committee on Citizenship and Immigration, 1 March 2001, 09:10.
48 *Ibid.*, 09:15–09:25.
49 Razack, 'Making Canada White'. See also Razack, 'Simple Logic', pp. 181–209.

50 Legislative Committee. E. Caplan, Standing Committee on Citizenship and Immigration, 1 March 2001, 09:10.
51 Legislative Committee. J. Fontana, Standing Committee on Citizenship and Immigration, 1 March 2001, 09:08.
52 Legislative Committee. S. Day, Standing Committee on Citizenship and Immigration, 1 March 2001, 09:30.
53 Goldberg, *The Racial State*, p. 257.
54 *Ibid.*
55 Razack, 'Making Canada White', p. 172.
56 See Raimondo, 'Corralling the AIDS Virus'; S. Watney, *Policing Desire: Pornography, AIDS, and the Media* (Minneapolis: University of Minnesota Press, 1987).
57 Legislative. E. Caplan, Standing Committee on Citizenship and Immigration, 1 March 2001, 10:25.
58 *Ibid.*
59 Legislative Committee. J. Atkinson, Standing Committee on Citizenship and Immigration, 13 March 2001, 10:45. My emphasis.
60 For a discussion of the international response to HIV/AIDS, see Klein, *HIV/AIDS*, pp. 27–31.
61 Balibar, *We the People*, p. 224.
62 Legislative Committee. E. Caplan, Standing Committee on Citizenship and Immigration, 1 March 2001, 10:25.
63 Legislative Committee. M. Battista, Standing Committee on Citizenship and Immigration, 22 March 2001, 09:50.
64 M. A. Somerville and S. Wilson, 'Crossing Boundaries: Travel, immigration, human rights and AIDS', *McGill Law Journal*, 43 (1998): 810.
65 Legislative Committee. J. Atkinson Standing Committee on Citizenship and Immigration, 13 March 2001, 10:50.
66 Somerville and Wilson, 'Crossing Boundaries', p. 805.
67 Legislative Committee. J. Atkinson, Standing Committee on Citizenship and Immigration, 5 April 2001, 10:55.
68 Legislative Committee. A. Klein, Standing Committee on Citizenship and Immigration, 5 February 2002, 10:45.
69 Legislative Committee. J. Atkinson, Standing Committee on Citizenship and Immigration, 13 March 2001, 10:45.
70 S.L. Gilman, *Disease and Representation: Images of Illness From Madness to AIDS* (New York: Cornell University Press, 1988), p. 7.
71 I have borrowed the term 'racial state' from Goldberg's book, *The Racial State*.
72 J. Hill, Weekly Column, www.jayhillmp.com (accessed 19 July 1995).
73 'Canada Reverses Immigrant AIDS Policy', *BBC News*. www.bbcnews.com (accessed 13 June 2001).
74 See C. Patton, *Inventing AIDS*, especially chapter 4; Watney, *Policing Desire*.
75 D.T. Goldberg, *Racist Culture: Philosophy and the Politics of Meaning* (Cambridge: Blackwell, 1993), p. 3.
76 W. Lubiano, 'Black Ladies, Welfare Queens, and State Minstrels: Ideological War by Narrative Means', in T. Morrison (ed.), *Race-ing Justice Engendering Power* (New York: Pantheon, 1992), pp. 332–3.
77 Razack, '"Simple Logic"', p. 189.
78 *Ibid.*, p. 195.
79 Lubiano, 'Black Ladies', p. 338.

80 M. Habib, 'Canada Boosts Funding to Fight HIV and AIDS', *The Canadian Press* (1 June 2000).
81 Patton, *Globalizing AIDS*, p. 27.
82 *Ibid.*, p. 32.
83 L. Priest, 'HIV Test Used to Bar Potential Immigrants', www.globeandmail.com (accessed 24 February 2003).
84 Priest, 'HIV Test'.
85 Watney, *Policing Desire*, p. 48.
86 *Ibid.*, p. 49.
87 Patton, *Globalizing AIDS*, p. 31.
88 'Coming to Canada with Dreams and HIV', *Globe and Mail* (2 October 2004): A1.
89 *Ibid.*
90 Patton, *Inventing AIDS*, p. 89.
91 See *African Canadian Legal Clinic Brief to the Legislative Review Secretariat, Citizenship and Immigration Canada*, 2001. http://www.aclc.net/submissions/immigration_refugee_policy.html
92 Legislative Committee. J. Fontana, Standing Committee on Citizenship and Immigration, Wednesday 2 May 2001, 11:30. Cited in Murdocca, *Foreign Bodies*, p. 64.
93 Goldberg, *Racist Culture*, p. 7.
94 W. Lubiano, 'Introduction', in W. Lubiano (ed.), *The House that Race Built* (New York: Vintage Books, 1998), p. viii.
95 Goldberg, *The Racial State*, p. 236.
96 Patton, *Inventing AIDS*, p. 99.
97 Hardt and Negri, *Empire*, p. 136.

9

Passports and Pestilence: Migration, Security and Contemporary Border Control of Infectious Diseases

Richard Coker and Alan Ingram

Since the late 1990s, increasing attention has been paid to the broader societal implications of certain chronic infectious diseases, above all HIV/AIDS. In particular, connections have been drawn with ideas of international security, and these are now reflected in post-9/11 global policy discourse, as are questions of migration.[1] As it affects issues of disease, this discourse is driven by a relatively small number of linked strategic concerns. The first is that in an era of increasing interconnections, diseases may spread more easily from one region to another through travel and migration. The second is that over the longer term disease may threaten economic interests. The third is that, in undermining the social, political and economic fabric and through its disproportionate effects on governance and military institutions, disease may in some places have destabilizing implications for state sovereignty and international security over the medium to long term. For many analysts, this requires that the familiar concerns of state security be complemented with the imperative of human security, or the protection of individuals and populations from threats to their well-being and existence, regardless of their citizenship, and, perhaps, location. This is challenging in a world that is still fragmented geopolitically and highly unequal economically. As a result, the discourse on globalization, disease and security that has emerged is beset by tensions, particularly in relation to questions of economic integration and migration.

In this context, borders represent an enduring and familiar tool for the control of threats to security. The connections between disease, security and borders have long histories, which demonstrate that public health practices at borders are shaped by attempts to secure states, identities and social arrangements. They also show that when security is threatened, responses are moulded by combinations of self-interest, knowledge and fear.[2] As other chapters in this collection demonstrate, public health practices have therefore intertwined historically with those of territoriality, states and identities, and have collided with economic interests. Although their utility and efficacy are increasingly questioned under globalization, the symbolism of borders for national identity, sovereignty and control sustains their attractiveness to a range of political actors. Borders have therefore inevitably become a constitutive element of the new security discourse. On the one hand, this stresses the vulnerability of domestic populations to transnational threats, necessitating overseas intervention to solve problems at source. According to Gro Harlem Brundtland, 'a single microbial sea washes all of humankind. There are no health sanctuaries. Diseases cannot be kept out of even the richest of countries by rearguard defensive action'.[3] On the other, borders and the surveillance of domestic space are to be reinforced in order to keep out threatening Others. This in turn increases the likelihood that phenomena such as disease and terrorism will be conflated with each other and with migration.[4]

Because differences in epidemiological dynamics have correlates in political dynamics, a focus on chronic infectious disease provides instructive contrasts with the experience of SARS. First, although they are distinct, transmission mechanisms for HIV/AIDS, tuberculosis and malaria are in general shaped to a far greater extent by patterns of wealth and poverty. Second, their direct short-term effects tend to be much less visible in political processes; rather their social impact is cumulative. Third, the stresses they place on health systems, where access to treatment exists, are longer term. With SARS and similar events, the surge capacity of health services to respond to spikes in demand is a critical concern. With chronic disease in the West, the issue is more to do with demands on welfare states suffering under pressures for cost containment. However, each of these differences can be short circuited in media, political and popular imaginations by migration, where metaphors of swamp, deluge and flood are common. The recourse to border control can then appear rational in terms of politics, regardless of whether it is in the interests of public health.

In this chapter we explore connections between migration, security and sovereignty in order to illustrate challenges for public health. In particular, we suggest that the recurring and politically powerful motif of protection of domestic populations from a threatening outside world by means of border control poses difficulties for rational public health policy. We begin by outlining the links that have been made between chronic infectious disease, globalization, development and security, noting the limits of global responses. Second, we focus on responses to these issues by the European Union (EU), an emerging foreign and security policy actor and long-time donor of development aid. Third, we examine in more detail issues around tuberculosis and migration, with particular reference to UK policy debates. The central paradox in all of this is that an excessive focus on border control will ultimately undermine protection against global infectious chronic disease.

Infectious disease, development and security

James Der Derian suggests that the term security conventionally refers to 'a condition of being protected, free from danger, safety'. Yet, he argues, 'the unproblematical essence that is often attached to the term today does not stand up to even a cursory investigation'.[5] Security can be constructed in many different ways, but the crucial questions will always be, who is to be protected, and from what?

Brower and Chalk argue, in a report for the Rand Corporation that 'Statecentric models of security are ineffective at coping with issues, such as the spread of diseases that originate within sovereign borders, but have effects that are felt regionally and globally. Human security reflects the new challenges facing society in the twenty-first century. In this model, the primary object of security is the individual, not the state. As a result, an individual's security depends not only on the integrity of the state but also on the quality of that individual's life.'[6] Migration and disease are therefore challenging for models of foreign and domestic security that rely on stable notions of 'here', 'there', 'us' and 'them'. Jordan Kassalow has argued from a US perspective that where public health and global interdependence are concerned this produces increasing alignment between three versions of the national interest: narrow self-interest; enlightened self-interest; and global engagement.[7]

Under the notion of narrow self-interest, Kassalow suggests that resurgent and emerging infectious diseases pose a direct threat to Americans and that this threat is growing. Within this construct he suggests that 'increased trade and travel, population movements, and a shared food supply spread health risks across the globe and the socioeconomic spectrum. People are more mobile: 57 million Americans traveled abroad in 1998, and tourism now claims to be the world's largest industry, accounting for 11.7 per cent of global GDP in 1999. There are significant movements of populations in the other direction, too: 70,000 foreigners enter the United States every day, and the nation had 26.3 million foreign-born residents in 1998'. Kassalow illustrates the global reach of some infectious diseases by highlighting the fact that, in 1998, of the 18,266 cases of tuberculosis reported in the US, 41 per cent occurred in foreign-born people. Countries of origin contributing the highest number of cases include Mexico, the Philippines, Vietnam, China, and India. He also notes that the New York City tuberculosis epidemic of the 1980s and 1990s 'traced cases back to 91 countries.' Under the rubric of narrow self-interest, and whilst acknowledging the difficulties in predicting future risks, Kassalow suggests that 'with as many as 1.6 billion people predicted to travel abroad each year by 2020, a fast-moving new lethal disease, a catastrophic flu epidemic, or a drug-resistant "superbug" could abruptly increase the level of risk Americans face. Such "new" risks are precisely those that are most difficult to manage'.[8] In 2000 the US National Intelligence Council stated that emerging and resurgent infectious diseases would also complicate US and global security, and affect strategic and military interests.[9]

From the perspective of enlightened self-interest, Kassalow suggests that sovereign states have a greater interest in absolute gains rather than relative gains, and that this requires better global health and greater political stability. A range of reports have suggested that infectious diseases, in particular AIDS, are in many states hindering the transition of systems of government to democratic regimes, undermining civil society, and challenging the development of sound political structures. The WHO Commission on Macro-economics and Health (CMH) argued that 'the linkages of health to poverty reduction and to long-term economic growth are powerful, much stronger than is generally understood. The burden of disease in some low-income regions, especially sub-Saharan Africa, stands as a stark barrier to economic growth and therefore must be addressed frontally and centrally in any comprehensive development strategy. The AIDS pandemic represents a unique challenge of unprecedented urgency and intensity. This single epidemic can undermine Africa's development over the next generation.'[10]

Narratives such as these postulate vicious circles. AIDS in particular is contributing to declines in productivity through a collapse in the available workforce. This in turn leads to a reduction in revenues and a redirection of scarce resources away from education and infrastructure towards health and social care. Failing economies become isolated from global trading partners and investment falls further, heightening competition for scarce resources and the likelihood of political instability and conflict. In turn this leads to population movements and with this the potential spread of infectious diseases.

The third perspective Kassalow draws is that of 'global engagement: the good leader'. Under this rubric he appeals to 'moral solidarity', arguing that the US 'as a rich and dominant nation, bears some responsibility for problems faced by those beyond its borders'.[11] Given the daily scale of suffering from preventable and treatable infectious disease, therefore, the US is simply obliged to act.

Each of Kassalow's levels is reflected in global policy discourse. But to what extent has rhetorical commitment translated into necessary and sufficient collective action? In 2000, all 191 member states of the United Nations adopted the Millennium Declaration, from which were derived a set of eight Millennium Development Goals (MDGs), with the headline goal of halving world poverty by 2015 in relation to 1990 levels.[12] Goal 6 is to 'Combat HIV/AIDS, malaria and other major diseases'. Under this are two targets: to have halted and begun to reverse the spread of HIV/AIDS by 2015, and to have halted the incidence and spread of malaria and other major diseases, including tuberculosis.

Despite this global commitment, the original MDGs had two notable omissions. First, reference to equity, justice and rights were not sustained from the Declaration into the MDGs, and second, Goal 8: the partnership for development, lacked any metrics or accountability for donor countries.[13] While the second of these shortcomings has been addressed retrospectively to some extent, the first has not.

At the same time, the salience of HIV/AIDS, tuberculosis and malaria in global policy concern did not just reflect recognition of their severity. It was also emblematic of a shift in donor preferences towards vertical, disease oriented programs and away from systemic approaches to health. This chimed with the rethinking of the role of the state under the neoliberal economic and political orthodoxy of the 1980s and 1990s. Given dissatisfaction with existing global health mechanisms and the political attractiveness of creating a new one, the G8 at their 2000 Okinawa summit, accepting that additional resources were required, agreed to create a new partnership on HIV/AIDS, tuberculosis and malaria. This led to the creation of the Global Fund, which began operating in 2002. However the fund has been afflicted by underfunding since its inception.

The Bush administration has claimed international leadership in the fight against disease, and pledged $15 billion for the President's Emergency Plan for AIDS Relief (PEPFAR) over 5 years and $1 billion per annum, possibly rising to $5 billion annum, for the Millennium Challenge Account, which aims to pioneer a new approach to development, based on competitive selection of countries that 'rule justly, invest in their people, and encourage economic freedom'.[14] While impressive in scale, both initiatives largely bypass existing multilateral mechanisms, thus increasing fragmentation and malcoordination, operate with a selective group of countries, and come hedged with conditions likely to undermine their effectiveness in tackling the diseases of poverty. They are also located within a broader foreign policy agenda of economic integration that contains significant risks to public health. As with the Global Fund, the preferred PEPFAR mechanism of focusing on specific diseases and vertically-implemented programs risk introducing perverse incentives, de-stabilizing already fragile health economies, and, if not sustained, promoting drug resistance.[15] Finally, they may be at risk of co-option to narrow security and foreign policy agendas.[16]

Overseas Development Aid (ODA) for health has been increasing since the late 1990s (to $8.1 billion in 2002), but even with these new commitments, it will still fall short of what the CMH estimated is necessary to provide even a minimally effective response ($27 billion in 2007, $38 billion in 2015).[17] In February 2005, UNAIDS estimated a 3 year shortfall in funding of the expanded response to HIV/AIDS in low and middle-income countries of US$8 billion. While the UN Security Council has declared HIV/AIDS to be a threat to international peace and security, the implication of this shortfall would be that, 'In many nations, the epidemic's trajectory would be barely affected, if at all.'[18] More generally, current pledges will not meet the doubling of ODA that would, together with a raft of other measures taken by rich and poor countries alike, be necessary to meet agreed development goals.[19]

European responses

The challenges of global chronic infectious disease have emerged onto European agendas in particular ways. For example, they are mentioned in the European Security Strategy adopted by the European Council in 2003, as in other manifestations of the new security discourse.[20] Although migration is not discussed, it can be taken as read in the context of concerns about instability, insecurity and underdevelopment, and the increasing role that the EU has been playing as a vehicle for migration policy. However, the particular constellations of institutions and powers that has resulted from the process of European integration means that chronic infectious disease is not explicitly at the frontline of migration and border policy.

The process of European integration involves geopolitical and geoeconomic tensions. Within the EU, integration has meant the promotion of 'the four freedoms': freedom of movement for capital, goods, services and labor. This has proceeded in tandem with efforts to liberalize the global economy on terms favorable to Europe. At the same time, politically influential and strategically significant domestic interests have been shielded from international competition. However, the European-born population is ageing, and, in some countries, shrinking. This, together with violence and instability elsewhere in the world, is driving immigration to the EU, which politicians in turn seek to control and shape to their own agendas. However, while disease, migration and borders are present in the EU variant of the new security discourse, the nature of concrete policy responses is determined by the differential development of EU powers.

The centrality of health policy to electoral politics in all European countries has meant that control has largely been retained by states, and authority within European institutions is consequently relatively weak. Since the 1980s, however, the EU has developed a role in response to demands for increased food safety and consumer protection. Furthermore, there was growing evidence that existing surveillance and control systems, based on national structures, were inadequate in the face of outbreaks that crossed borders. Article 129 of the 1992 Maastricht Treaty provided a legal basis for action in this field, and led to a program that provided short-term project funding to networks assembled largely by groups of enthusiasts in national surveillance centers and academic departments.[21] A new European Center for Disease Prevention and Control began work in May 2005. Much of the impetus for the new center has come from SARS and fears of bioterrorism, along with worsening public health in Eastern Europe and the former Soviet Union. The new organization is, however, very much smaller than its equivalent in the US. It aims to link existing surveillance and laboratories, provide early warning and response, and offer scientific opinions and technical assistance.

In contrast to its limited role in public health policy, the EU has long been significant in international development, and has expressed commitment to achievement of the MDGs and the Cairo goal of universal sexual and reproductive health and rights, both of which have direct relevance for chronic infectious disease. Because privileged trading relationships with European colonies and post-colonial successor states were affected by the single market, the European Commission gained significant competence in development aid early in the process of European integration. But this has in general not been well focused on the poorest countries: much aid has been tied to the purchase of goods and services from donor countries, and administration has been hugely inefficient. Since the end of the Cold War, the balance of aid has shifted away from the developing world (albeit in large part the less poor regions) and towards the 'near abroad'; that is, the Western Balkans, Eastern Europe, and former Soviet Union, regions which do not qualify for ODA. The harmful effects of the Common Agricultural Policy and certain other trade barriers on international development, have also received widespread condemnation.[22]

As a result of severe criticism from developing countries, NGOs and certain donor countries (particularly the UK), the EU has embarked on reform, oriented around the Millennium Development Goals.[23] The Commission has also been active in international policy debates on HIV/AIDS, pledging to make good the funding deficit of the UN Population Fund (UNFPA) after the Bush administration withdrew its support as its first legislative act in 2001. But while EU aid contains a significant focus on the diseases of poverty, the proper balance between poor countries (where the impacts of chronic infectious disease are greatest) and the near abroad (which is home to worsening HIV/AIDS and tuberculosis epidemics) is as yet unresolved.

The EU has also been developing a role in migration and border policy, with authority in these fields delegated by member states in the Maastricht Treaty of 1992, following earlier intergovernmental cooperation on the Schengen area. It has been argued that because oversight of policy in these fields is relatively weak at the EU level, member states have used the EU as a vehicle to enact tougher policies than they might be able to justify at home. This is particularly evident in relation to asylum and 'illegal migration'.[24] Increasing linkages have also been made between development, security and migration policies in an attempt to ensure that people from poor and conflict-affected regions remain outside the EU or can be returned there.[25] Indeed, the drive for greater coherence of EU policies around security concerns has given rise to trepidation about a narrowing of priorities just as broader human security agendas are beginning to emerge.

While the screening of migrants to the EU for chronic infectious disease is not an explicit part of EU policy, for reasons to do with the politics of European integration, people with these diseases or from regions with high rates of diseases like tuberculosis and HIV/AIDS are likely to fall within the broader exclusionary approach to migration and borders being implemented through the EU itself.

UK responses

Chronic infectious diseases, above all HIV/AIDS and tuberculosis have been much more salient in UK migration debates than at the EU level. This sits within the broader discourse on globalization and security, which UK policy makers have helped to produce. But UK policy is distinguished by the extent to which policy makers have, in declaratory terms at least, begun to qualify the neo-liberal orthodoxy in international development, and the way they have built legislative and institutional firewalls between development and security. This means that chronic infectious diseases are specifically framed as worthy of concern in terms of humanitarianism and development rather than security.[26] However, the salience of migration and the NHS as frontline electoral issues mean that different standards may be applied to sections of the global poor and diseased should they arrive in the UK itself, and that public health principles may be traded off with political imperatives. Finally, the domestic policy priority of recruiting large numbers of health professionals from other (inevitably poorer) countries conflicts with the goal of supporting health systems development.

Tony Blair has made clear his commitment to the struggle against HIV/AIDS, as has the Chancellor of the Exchequer, Gordon Brown.[27] A rising proportion of the aid budget has been allocated to tackling the pandemic, and while much of this money is channeled bilaterally, the UK has been a major supporter of UNAIDS and the Global Fund, and of increasing access to medicines. Together, these policies will make a substantial contribution to increasing the number of people in poor countries who have access to treatment. This can be set more generally in the context of commitment to the MDGs and their adoption as an evaluation framework for UK policy on international development. The UK has also used its considerable policy networks and influence to encourage international organizations such as the WHO and the EU to adopt the MDGs in the same way. The UK has thus helped to pioneer what Simon Maxwell has called the new meta-narrative in international development.[28]

Some of this activity can be attributed to commitments the New Labour administration brought to power in 1997. The new government immediately established a Department for International Development (DFID) independent from the Foreign and Commonwealth Office (FCO), and defined the purpose of development assistance as the reduction of poverty in legislation in 2002.[29] Under further pressure from transnational networks of NGOs, churches and trade unions, as well as developing countries themselves, the UK has also begun to qualify the commitment to neo-liberalism that is still dominant domestically. In March 2005, DFID, the FCO, and the Treasury announced that they would no longer insist on privatization as a condition of UK aid, but would focus rather on outcomes in poverty reduction, and urged other donors to do the same.[30] Finally, the government has reaffirmed that security and development, while related, are two distinct policy fields, and that the latter will not be subordinated to the former.[31]

Globally, then, the strategy is to tackle the diseases of poverty, in part by expanding access to treatment. The picture at home is rather different. Two concerns have dominated. First, there have been calls for expanded screening of immigrants to the UK to protect public health and the public purse, and second, steps have been taken to deny access to the NHS by certain categories of people. Here it is not the security of the state that is said to be threatened, but the terms of UK citizenship, with border controls functioning as a form of defense.

For much of the 1990s, opinion polls tended to show that less than 10 per cent of people in the UK viewed immigration and asylum as among the most important issues facing the country.[32] This proportion began to rise during 2000, and reached a peak of 40 per cent in early 2005, driven by increasing immigration and asylum applications, and the politicization of these issues by the media, political parties and advocacy groups. During this period, polls also tended to find that the NHS/hospitals were most often cited as the most important issue. Given the shift in the burden of tuberculosis and HIV to people born outside the UK during the same period, the emergence of chronic infectious disease onto the political agenda is easy to explain. The political salience of chronic diseases peaked in 2003, during a period of heightened anxiety about asylum and immigration, and again in 2005, in the run up to the General Election in May of that year.

Calls to expand screening of immigrants to the UK for disease were made by the opposition Conservative Party in 2003 and repeated in early 2005.[33] While rejecting these policies in general, the government announced that it would introduce screening for tuberculosis for potential migrants from 'high-risk countries'. As we discuss below in relation to tuberculosis, the evidence base, practicality, and ethical viability of these proposed policies are very much open to question.

Second, responding to anxieties about 'health tourism' and 'health asylum' fueled by some politicians and sections of the media, in April 2004 the Department of Health restricted free access to treatment for

HIV/AIDS for certain categories of people that have 'no substantive connection to the UK'.[34] Although diagnostic testing and counseling for HIV/AIDS remained free, charges were introduced for some people who require treatment. These included asylum seekers whose claims have been rejected and those who have not been lawfully resident in the UK for 12 months. The onus of enforcement fell on the health care provider; the new regulations stated that 'where a patient is found to be HIV positive and treatment, including drugs, is needed, the hospital or sexually transmitted diseases clinic need to have systems in place to establish if that patient is ordinarily resident or, if not, exempt from charges.' This policy was introduced without proper evidence, and despite serious concerns about its public health effects, economic costs, administrative feasibility, and ethical implications.[35]

The divergence between global and domestic policies on chronic infectious disease has economic, moral and political dimensions. In poor countries, HIV/AIDS is seen as a barrier to economic development; in the UK, it is a costly condition that creates a drain on public finances. In poor countries, HIV/AIDS is a humanitarian disaster; in the UK, the NHS 'is there to provide free treatment for those who live here and not for those who do not'.[36] Politically, there is strong support for international development within the Labor Party and the public at large. Consequently political advantages flow from being seen to act as a good international citizen; domestically, political advantages accrue from appearing to be tough on issues such as asylum and perceived 'scrounging', and to protect the NHS. This led to concerns about the negative effects of the UK immigration and nationality system on people living with HIV, and their implications for UK public health.[37]

Much of the debate about these issues has been characterized by a distinct lack of evidence and a narrow conception of globalization. While some organizations and individuals have been vocal in expressing concerns about illegitimate use of the NHS, the extent of misuse by overseas visitors and asylum seekers has not been quantified systematically. Neither have public and political debates set the cost of caring for people from overseas against the benefits that the UK public and private sectors, and the NHS in particular, reap from greater international mobility. This has often been at the expense of poor countries whose investment in training of health professionals provides a subsidy to the UK taxpayer. The global total subsidy to rich countries represented by migration of health professionals has been estimated at $500 million per year, with the UK undoubtedly capturing a significant proportion.[38] Despite clear evidence of the harmful effects of the brain drain on the health systems of poor countries, the policy response has to date been fragmented, limited and ineffective, when the need for a comprehensive approach has been established.

Tuberculosis and migration in the UK

Many of the issues raised in this chapter are illustrated vividly by the policy debate over tuberculosis control, migration and borders in the UK, which has been driven by domestic political imperatives, rather than the public health evidence base. When the WHO drew attention to the 'global emergency' of tuberculosis in 1993, it marked the re-emergence onto the radar screens of policy makers, of a disease that had never really gone away. Rather, tuberculosis was simply taking advantage of changes in patterns of poverty, overcrowding, and malnutrition as it always has. It was also taking advantage of other, newer, forces. HIV/AIDS has transformed the transmission dynamics of tuberculosis, and health systems are left struggling in the wake of this epidemiological onslaught. In short, the HIV/AIDS pandemic fuels an epidemic of tuberculosis, and previously successful control programs are being rendered ineffective by the emergence of resistant strains. Africa is the one major world region where rates of tuberculosis continue to increase.[39]

Tuberculosis has also taken advantage of socio-political turmoil. With fractured health systems, many states of the former Soviet Union have witnessed increasing rates of tuberculosis and anti-microbial resistance. The geographical re-alignment of the European Union in May 2004 with the accession of ten new member states means that the Union's new border to the east abuts states with worse infectious disease profiles. Notification rates in Western Europe overall decreased by 3.6 per cent yearly between 1995 and 2001, and in seven countries in Central Europe, by 4–6 per cent yearly (with rises in Bulgaria and Romania). In the former Soviet Union, rates in 2001 were 62 per cent higher than in 1995, with mean annual increases of 6–12 per cent in most countries.[40] The dynamics of the disease also vary from West to East; in 2001, whilst tuberculosis amongst foreign-born persons or non-citizens accounted for 32 per cent of all tuberculosis cases in Western Europe, only 1 per cent of cases in both Central and Eastern Europe were in foreign-born individuals.

The scourge of tuberculosis has also remained a persistent problem in deprived neighborhoods in affluent Western states, amongst largely disenfranchised populations such as the homeless, drug users, ethnic minorities, and migrants. The epidemic of multidrug-resistant tuberculosis in New York City in the late 1980s and early 1990s in particular caught the attention of local, national and international policy makers. A successful public health response refuted the earlier nihilistic despondency of some commentators. However, action on the local and global upstream fractures in health and social systems that helped to generate the epidemic has in general been less effective.[41] This means that tuberculosis remains a transborder health issue.

Notification rates for tuberculosis in England and Wales have been rising year-on-year since the mid-1980s.[42] These increases have disproportionately occurred in London, and the proportion of cases of tuberculosis occurring in people born abroad has been increasing too. For example, in 1988, 45 per cent of cases of tuberculosis were in people born abroad; by 2002, this proportion had increased to 67 per cent. Tuberculosis also occurs disproportionately in individuals from ethnic minorities compared to the white population. In 2002, rates in black Africans were around 280 per 100,000 population compared to 3.6 per 100,000 in the white population. Among black Africans born abroad, the rate was around 366 per 100,000. There is therefore little question that the burden of tuberculosis in England and Wales is associated with individuals born abroad. While it remains somewhat unclear whether tuberculosis is acquired whilst abroad or whilst resident in the UK, a reasonable assumption is that most disease in the foreign-born is probably acquired in the country of birth, prior to entry to the UK.

The question of screening immigrants and asylum seekers for tuberculosis has been a central theme in recent UK debates on migration and disease. Early in 2003, it was reported that a Cabinet-level Working Group on Imported Infectious was considering policy options, including compulsory screening of immigrants and asylum seekers in countries of origin.[43] As Convery, Welshman and Bashford discuss in their chapter, in August the opposition Conservative Party proposed an approach similar to that taken in Australia: before individuals could be given permission to remain in the UK, three tests would have to be met: they must not pose a risk of transmitting an infectious disease to the public; they must not create undue demand on restricted health resources; and they must not create a long-term drain on the public purse. The proposals suggested that 'those entering the UK through the immigration system would require (sic) to have such tests at the point of application and to pay for them, while those seeking asylum would be detained until it was clear the criteria had been met'.[44] This amounted to pre-entry screening for all permanent migrants to the UK and compulsory on-entry testing (and in this case, detention) for all asylum seekers. These proposals were not adopted by the Government, but screening remained close to the surface of the UK political agenda through to 2005.

Many states do in fact screen immigrants for HIV and active tuberculosis, and this has been part of the political justification advanced by its advocates.[45] Indeed, the WHO advocates, in a model legislative framework, that when 'crucial' to public health, 'the population, or particular groups of it shall have a duty to undergo X-ray examinations, tuberculin tests, blood tests, or other comparable tests that can be carried out without danger'.[46] More specifically, the WHO advances that:

> Foreign-born persons intending to stay in the country (other than for a stated short period of time, for example not more than three months),

> who are not exempt from any residential permit requirement, have a duty to undergo medical examination for tuberculosis. If the foreign-born person is 15 years of age or more, he or she also has a duty to have a chest X-ray...The examination must be carried out as soon as possible and no later than three months after entry into the country. Refugees, asylum seekers and persons applying for a residential permit for the purpose of reuniting with their families must be examined within 14 days after entry.[47]

Despite such recommendations, service practices across Europe lack coherence. Whilst approximately half of countries in the pre-May 2004 European Union have no policies, those that do (broadly speaking, those in Western Europe) have policies that screen varying populations, with varying tools, that are interpreted differently.[48] The likely reason for this is that considerable uncertainty still exists regarding what is the most effective means of screening.

There are three major areas of concern about the practical efficacy of screening. First, the tools used to detect individuals with disease lack discrimination – the insensitivity of screening tools such as chest X-ray means that, as the prevalence of disease falls in the population screened, the number of false positives rises substantially (with important resource implications).[49] Second, immigrants with disease probably do not pose a substantial public health threat through delays in seeking care. Delays in diagnosis seem to be more of an issue in white populations, and especially women. This means that a particular focus on those from 'high risk' areas may be misplaced.[50] Moreover, delays that do occur can be traced to service weaknesses as well as patient-centered factors.[51] Third, much of the debate around screening new entrants for tuberculosis has centered on the assumption that, if done more effectively, benefits might accrue. But most foreign-born individuals who go on to develop tuberculosis do so some time *after* arrival – that is, very few enter the country with active disease. In February 2005, it was announced that screening of 150,000 immigrants at Heathrow airport resulted in the identification of only 100 cases of active tuberculosis, against an expected rate of 150 per 100,000.[52] Figures released by the Government showed that a project to screen asylum seekers in Kent in 2004 and 2005 found that only nine people out of 4,219 (or 0.2 per cent) had the active disease.[53] Modeling studies have also suggested that even if coverage through screening of new entrants was wider, very few cases of tuberculosis would be detected.[54]

There also tends to be an assumption that people arriving in the UK carry with them a risk of developing tuberculosis that corresponds to the prevalence of the country from which they came. This assumption is embedded in the notion that screening for tuberculosis should be focused upon those coming from countries where tuberculosis is considered 'common'.[55] Yet, research findings suggest there is in fact little correlation between the prevalence of tuberculosis in countries from which people originate and their risk of active disease on entry to the UK.[56] Nonetheless, in its 5-year strategy for immigration and asylum (in effect manifesto commitments for a third Labor term in office), which were designed to counter opposition Conservative criticisms of official policy in the forthcoming election campaign, the Government proposed to screen 250,000 visa applicants for tuberculosis 'on high risk routes' and require those diagnosed to seek treatment before being allowed to travel to the UK.[57]

Although such a policy might be appealing because of the potential to shift cost and responsibility (in line with the general move to contain potential immigrants and asylum seekers in their regions and countries of origin and buffer zones around the EU), and in terms of the political messages it sends ('our borders are secure, we are keeping disease at bay'), it is unlikely to be effective, and the diversion of resources will have opportunity costs that may undermine effective public health policy in the UK and poor countries. By defending the country from disease in this way, UK public health may in fact become less secure.

Conclusion

People are increasingly on the move, and migration flows are becoming more diverse and complex. Patterns of disease are also shifting in response to social, economic and political change. At the same time, immigration, asylum and border control have, like many fields of official policy, experienced a wave of securitization. Much of the official discourse on disease and security focuses on acute infectious diseases like SARS and those produced by biological weapons, which spread relatively rapidly and, in the case of the latter, may have, or be intended to have, direct political effects. Chronic infectious diseases have also been increasingly linked to state and international security concerns. There have therefore been high-level initiatives to address HIV/AIDS, tuberculosis and malaria, but at current rates of progress these seem unlikely to lead to global control in the near future. In this context, international migration and mobility will continue to bring home chronic infectious disease for rich countries.

With domestic health policy in rich countries dominated by debates about the allocation of scarce resources, it is the security of welfare states, linked to ideas of citizenship, that are to the fore, rather than the state itself. While globally the focus is on expanding access to treatment, locally the debate is heavily influenced by the politics of migration and the notions of 'deserving', 'undeserving', self and other that go with it. Also to the fore is the image of the nation at risk from a threatening external world. Hence, screening policies, aside from any public health merit, help to reinforce in political discourse the image of the effective state that can control flows across frontiers and keep at bay threats to health and the social order.

Borders can be seen as faultlines and frontlines in a fragmented world. As official concerns about migration, security and disease have increased, border controls have moved to the center of political debate. The construction of coherent, effective and just public health policies in this context is always likely to be challenging.

Notes

The work Richard Coker conducted for this chapter was undertaken before being seconded to the UK Department of Health. He did not contribute after his secondment, and the views expressed do not reflect those of any government department.

1 G.W. Bush, *National Security Strategy of the United States of America* (Washington DC: The White House, 2002), http://www.whitehouse.gov/nsc/nss.pdf (accessed 28 March 2005); A. Ingram, 'The New Geopolitics of Disease: Between global health and global security', *Geopolitics*, 10 (2005): 522–45.

2 A. Kraut, *Silent Travellers: Germs, genes and the 'immigrant menace'* (New York: Basic Books, 1994); D. Porter, *Health, Civilization and the State: A history of public health from ancient to modern times* (London: Routledge, 1999); A. Bashford and C. Hooker, 'Contagion, modernity and postmodernity' in A. Bashford and C. Hooker (eds) *Contagion: Historical and cultural studies* (London: Routledge, 2001), pp. 1–12; H. Markel and A.M. Stern, 'The Foreignness of Germs: The persistent conflation of immigrants and disease in American society', *Milbank Quarterly*, 80 (2002): 757–88; J. Collin and K. Lee, *Globalisation and Transborder Health Risk in the UK: Case studies in tobacco control and population mobility* (London: the Nuffield Trust, 2003), chapter 2.

3 G. Harlem Brundtland, 'Health and Population', Reith Lectures 2000, Number 4, http://www.bbc.co.uk/radio4/reith2000/lecture4_print.shtml (accessed 21 April 2005).

4 As expressed, for example, in relation to the US-Mexican border by E. Nelson, 'The Next American Revolution', http://www.usbc.org (accessed 18 April 2005).

5 J. Der Derian, 'The Value of Security' in D. Lipschutz (ed.), *On Security* (New York: Columbia University Press, 1995), p. 28.

6 J. Brower and P. Chalk, *The global threat of new and reemerging infectious diseases: reconciling US national security and public policy* (Washington, DC: Rand Corporation, 2003), p. xiii.

7 J.S. Kassalow, *Why health is important to US foreign policy* (Milbank Memorial Fund: New York, 2001), http://www.milbank.org/reports/Foreignpolicy.html (accessed 13 July 2005).

8 Kassalow, *Why Health Is Important*, online.

9 National Intelligence Council (2000) *The Global Infectious Disease Threat and its Implications for the United States* (Washington DC: NIC).

10 Commission on Macroeconomics and Health, *Macroeconomics and Health: Investing in health for economic development* (Geneva: WHO, 2001), pp. 1–2.

11 Kassalow, *Why Health Is Important*, online.

12 UN Millennium Development Goals, http://www.un.org/millenniumgoals/ (accessed 21 April 2005).

13 S. Maxwell, 'The Washington Consensus is Dead: Long live the meta-narrative!' *ODI Working Paper 243* (London: Overseas Development Institute, 2005).
14 Office of the United States Global AIDS Coordinator, *The President's Emergency Plan for AIDS Relief: US Five-Year Global AIDS Strategy* (Washington DC: Department of State, 2004), http://www.state.gov/documents/organization/29831.pdf (accessed 21 April 2005); Millennium Challenge Corporation, 'About the Millennium Challenge Account', http://www.mca.gov/about_us/overview/index.shtml (accessed 21 April 2005).
15 P. Travis et al., 'Overcoming health-system constraints to achieve the Millenium Development Goals', *The Lancet*, 364 (2004): 900–6.
16 S. Radelet, 'Bush and Foreign Aid', *Foreign Affairs*, September/October 2003, pp. 104–17. http://www.cgdev.org/docs/Bush_and_Foreign_Aid.pdf (accessed 26 March 2005).
17 R. Labonte, T. Schrecker and A.S. Gupta, 'A global health equity agenda for the G8 summit', *British Medical Journal*, 330 (March 2005): 533–6.
18 UNAIDS, *Resource Needs for an Expanded Response to HIV/AIDS in Low and Middle Income Countries* (Geneva: UNAIDS, 2005), pp. 2, 26. http://www.dfid.gov.uk/news/files/aidsresources9mar05.pdf (accessed 26 March 2005).
19 Maxwell, 'The Washington Consensus is Dead'.
20 J. Solana, *A Secure Europe in a Better World: A European security strategy* (Brussels: European Council, 2003), http://ue.eu.int/uedocs/cmsUpload/78367.pdf (accessed 21 April 2005).
21 L. Maclehose, R. Coker and M. McKee, 'Communicable disease control: detecting and managing communicable disease outbreaks across borders' in M. McKee, L. Maclehose and E. Nolte (eds) *Health policy and European Union enlargement* (Maidenhead: Open University Press, 2004).
22 S. Maxwell and P. Engel, *European Development Cooperation to 2010* (London: Overseas Development Institute, 2003).
23 Commission of the European Communities, *EU Report on Millennium Development Goals 2000–2004: EU contribution to the review of the MDGs at the UN 2005 High Level Event* (Brussels: European Commission, 2005), http://europa.eu.int/comm/development/body/communications/docs/132_com_staff_working_doc_en.pdf#zoom=100 (accessed 21 April 2005).
24 L. Fekete, *The Deportation Machine: Europe, asylum and human rights*, Introduction (London: Institute of Race Relations, 2005), http://www.irr.org.uk/2005/april/ha000011.html (accessed 26 April 2005); M. Samers, 'An emerging geopolitics of "illegal" immigration in the European Union', *European Journal of Migration and Law*, 6 (2004): 23–41.
25 B. Hayes and T. Bunyan, 'Migration, Development and the EU Security Agenda' in H. Mollett (ed.), *Europe in the World: Essays on EU foreign, security and development policies* (London: BOND, 2003), pp. 71–80.
26 A. Ingram, 'Global Leadership and Global Health: Contending meta-narratives, divergent responses, fatal consequences', *International Relations*, 19 (2005): 381–402.
27 T. Blair, 'PM marks World AIDS day', 1 December 2004, http://www.number-10.gov.uk/output/Page6695.asp (accessed 14 January 2005); G. Brown, 'A comprehensive plan for HIV/AIDS', http://www.hm-treasury.gov.uk (accessed 12 January 2005).
28 Maxwell, 'The Washington Consensus is Dead'.
29 Department for International Development, *International Development Act* (London: HMSO, 2002), http://www.legislation.hmso.gov.uk/acts/acts2002/20020001.htm (accessed 21 April 2005).

30 Department for International Development, *Partnerships for Poverty Reduction: Rethinking conditionality* (London: DFID, 2005), http://www.dfid.gov.uk/pubs/files/conditionality.pdf (accessed 27 March 2005).
31 Department for International Development, *Fighting Poverty to Build a Safer World: A strategy for security and development* (London: DFID, 2005), http://www.dfid.gov.uk/pubs/files/hivaidstakingaction.pdf (accessed 28 March 2005).
32 Market Opinion and Research International, *MORI Political Monitor: Long term trends*, http://www.mori.com/polls/trends/issues.shtml (accessed 27 April 2005).
33 Conservative Party, *Before it's too late: A new agenda for public health* (London: Conservative Party, 2003).
34 J. Hutton Foreword to *Implementing the Overseas Visitors Hospital Charging Regulation: Guidance for NHS Trusts in England* (London: Department of Health, 2004), p. iii., http://www.dh.gov.uk/assetRoot/04/10/60/24/04106024.pdf (accessed 17 January 2005). (Plans to introduce compulsory HIV tests for immigrants were shelved. J. Revill, 'Ministers drop HIV test plan', *The Observer*, 25 July 2004, http://observer.guardian.co.uk/politics/story/0,,1268811,00.html (accessed 27 April 2005).
35 Refugee Council 2004. Prime Minister announces plans to step up asylum removals (16 September 2004) http://www.refugeecouncil.org.uk/news/sept04/relea178.htm (accessed 14 January 2005); B. Gazzard et al. *Treat With Respect: HIV, public health and immigration*. Report of expert panel, 2005, http://www.irr.org.uk/pdf/HIV_Treat_With_Respect.pdf (accessed 25 April 2005).
36 Hutton, 'Foreword'.
37 All-Party Parliamentary Group on AIDS, *Migration and HIV: Improving lives in Britain* (London: APPG Aids, 2003) http://www.appg-aids.org.uk/Publications/Migration%20and%20HIV%20Improving%20Lives.pdf (accessed 27 April 2005).
38 Joint Learning Initiative, *Human Resources for Health: Overcoming the Crisis* (Cambridge, Mass: Harvard University Press, 2004), p. 102. http://www.globalhealthtrust.org/report/chapter4.pdf (accessed 17 January 2005).
39 World Health Organization, *Global Tuberculosis Control: Surveillance, planning, financing* (Geneva: WHO, 2005).
40 EuroTB (InVS/KNCV) and the national coordinators for tuberculosis surveillance in the WHO European Region, *Surveillance of Tuberculosis in Europe: Report on tuberculosis cases notified in 2001* (Saint-Maurice: Institute de vielle sanitaire, 2003) http://www.eurotb.org/rapports/2001/etb_2001_full_report.pdf (accessed 25 April 2005).
41 A.L. Fairchild and G.M. Oppenheimer, 'Public health nihilism versus pragmatism: history, politics, and the control of tuberculosis', *American Journal of Public Health*, 88, 7 (1998): 1–14. R.J. Coker, *From chaos to coercion: detention and the control of tuberculosis* (New York: St. Martin's Press, 2000). S.N. Tesh, *Hidden Arguments* (New Brunswick and London: Rutgers University Press, 1988).
42 R. Coker, *Migration, public health and compulsory screening for TB and HIV. Asylum and Migration Working Paper 1* (London: Institute for Public Policy Research, 2003).
43 A. Travis, 'Asylum seekers may face health checks', *Guardian Unlimited* (24 November 2003), http://www.guardian.co.uk/uk_news/story/0,,1091757,00.html (accessed 27 April 2005).
44 Conservative Party, *Before it's too late.*

45 M. Foreman et al., *The third epidemic: repercussions of the fear of AIDS* (London: Panos Institute and Norwegian Red Cross, 1990). A.C. Hayward et al., 'Epidemiology and control of tuberculosis in western European cities', *International Journal of Tuberculosis and Lung Disease*, 7 (2003): 751–7.

46 G. Pinet, *Good practice in legislation and regulations for TB control: an indicator of political will* (Geneva: World Health Organization, 2001).

47 *Ibid.*, p. 54.

48 R. Coker, A. Bell, R. Pitman, J. Watson, 'Screening programmes for tuberculosis in new entrants across Europe', *International Journal of Tuberculosis and Lung Disease*, 8 (2004): 1022–6.

49 Coker, *Migration, public health and compulsory screening for TB and HIV.*

50 A. Rodger et al., 'Delay in the diagnosis of pulmonary tuberculosis, London, 1998–2000: analysis of surveillance data', *British Medical Journal*, 326 (2003): 909–10.

51 S. Paynter et al., 'Patient and health service delays in initiating treatment for patients with pulmonary tuberculosis: retrospective cohort study', *International Journal of Tuberculosis and Lung Disease*, 8 (2004): 180–5.

52 M. Tempest, 'Tory immigrant screening plan "chaotic"', *Guardian Unlimited* (15 February 2005), http://politics.guardian.co.uk/homeaffairs/story/0,,1415054,00.html. (accessed 26 April 2005).

53 D. Batty, '0.2% of asylum seekers have TB', *Guardian Unlimited* (17 February 2005), http://www.guardian.co.uk/uk_news/story/0,,1416708,00.html (accessed 27 April 2005).

54 R. Pitman et al., *Modelling new entrant screening for tuberculosis in England and Wales*, Poster abstract, HPA Conference, Warwick, 13–15 September, 2004.

55 The prevalence of 40/100,000 is the 'arbitrary but reasonable level above which tuberculosis may be considered "common"' according to the British Thoracic Society and the prevalence rate at which screening in the UK is currently advocated.

56 R. Coker, 'Compulsory screening of immigrants for tuberculosis and HIV', *British Medical Journal*, 328 (2004): 298–300.

57 Home Office, *Controlling Our Borders: Making migration work for Britain* (Norwich: HMSO, 2005). 'Tory immigrant screening plan'.

Part III

Globalization: Deterritorialized Health?

10

Drawing the Lines: Danger and Risk in the Age of SARS

Claire Hooker

'It's really just a question of where people are going to draw the lines'.
Dr Alison McGreer, 2003[1]

'We are living in a new normal ... The old days where an infection might emerge every now and again and capture our attention really has changed'.
Dr Julie Gerberding, Director, US Centers for Disease Control, 2004[2]

The outbreak of SARS in the spring of 2003 was shocking for many reasons. It seemed to realize the threat that had been haunting public health professionals' imaginations: that a new or re-emerging infectious disease could wreak havoc even in Western nations. And the havoc was measured in more than just mortality. The outbreak saw the resurrection of instruments long out of use, such as containment by quarantine and 'social distance' measures, and their social and economic costs were devastating for the healthy people who had to bear them.

The SARS epidemic had the feeling of a watershed, a marker of historical change: on one hand, a return to an unhappy past when infectious disease threatened Western nations that had felt free of epidemics, and on the other, an optimistic transition to a global public health with the potential for swiftly identifying and containing in a place of origin any disease that might prove a threat to all. Public health scientists in Canada exclaimed over the terrors of using quarantine, an ancient infection-control instrument with which most had no direct familiarity, and simultaneously over the unprecedented international cooperation that produced usable data on the coronavirus in record speed. In this situation, where new and old intermingle, I examine the response to the outbreak of SARS in Toronto through the theoretical lens that has viewed changing public health instrumentalities in the past: the transition in twentieth-century public health from instrumentalities based on a logic of 'dangerousness' to those based on 'risk'.[3] I argue that the response to the SARS outbreak was an irruption of the logic of dangerousness into risk-based public health. Containment strategies such as quarantine, policing air travel and hospital closures, which had significant negative social and economic impacts, resulted from a logic of finding and removing danger. But since many public health instruments have the capacity to work on the basis of either logic, I argue that the difference between dangerousness and risk is better understood as a scientific and policy *stance* in which, effectively, a choice between consequences is being made. Whether one quarantined newly-adopted Chinese children or only nurses in hospitals with SARS patients,[4] whether one spent money on thermal scanners in airports or on hospital emergency wards, such decisions depended on where one chose to 'draw the line'.

From dangerousness to risk

The idea that there has been a shift in the basic logic of public health from dangerousness to risk was elaborated in an article by Robert Castel, who, taking up a suggestion by Michel Foucault, argued that until the mid to late twentieth century the central logic of public health governance (he was more narrowly interested in mental illness) was one of 'dangerousness'.[5] By this he meant that authorities were preoccupied with locating and neutralizing all sources of danger, that is all threats to health. The instruments used to do this were those with which historians of public health are familiar: chiefly, quarantine and isolation or incarceration, together with disinfection and whatever therapies were available at the time. But the strategy of locating and neutralizing every person, place or object that was dangerous was limited, firstly, because of the difficulties of locating the dangerous, who, prior to committing a crime or transmitting an illness, mostly showed no sign of the threat they posed, and secondly because once identified, they could only be dealt with one by one. Attempts to control venereal disease in the early twentieth century offer a good illustration of these limitations. How could infected people be located and prevented from infecting others, when they were physically asymptomatic and when the sexual behavior considered to be disease-causing was hidden? The best-known instrument of the time, the 'Lock Hospital',[6] where prostitutes were treated and incarcerated, was obviously ineffective, especially when male sexual partners were not similarly sanctioned.

Castel argues that in the second half of the twentieth century the logic of intervention changed. As new concepts and methods in epidemiology developed, such as those that grew out of the pioneering studies of the link between tobacco smoking and mortality,[7] threat was no longer considered to arise from the presence of a particular danger embodied in a concrete entity. Instead it was seen as the effect of abstract factors: in other words, a risk, one that could be more or less precisely calculated. In our example, the spread of sexually-transmitted diseases (the change in terminology matches the change in preventive instrumentalities) can be calculated from the combination of age, gender, level of education, background and so forth, and changing these factors can lower rates of disease without confining any particular individuals.

This move overcame the limitations of the logic of dangerousness. Rather than attempting to confront each concrete dangerous situation, experts calculated combinations of risk factors in the population and designed interventions for the resulting, newly 'identified' (more accurately, 'constructed') at-risk groups. Castel argued that in late twentieth-century preventive instrumentalities, intervention was based not in the interactions between medical practitioners and patients, but in the technocratic management of 'flows of population'. In our example, chlamydia would now be prevented by programs managing teenage girls: teenage girls with a specified level of education and a specified level of self-confidence and so on.

Rather than imposing constraints on people, the pre-eminent strategy of this 'new public health' was health promotion, a series of connected instruments (mass media campaigns, participation-action research, workplace regulation, health education and risk communication) that governed through people's autonomy. Each person, sorting through the barrage of information directed their way by authorities, is supposed to identify which factors place them at risk of ill-health, and take the actions recommended by health experts to avoid them.[8] (Parents of teenage girls should make them play sport, and teenage girls should watch the ads and consequently use condoms.) It has not been lost on Foucauldian scholars following Castel that the 'new public health' was valorized at the very time that public health funding (and, as a result, infrastructure) were cut in the entrepreneurial 80s and pragmatic 90s – a fact of no small significance in the SARS story.[9] Effectively, social destinies are assigned through resource allocation by administrators in relation to individuals' varying capacity to live up to the requirements of competitiveness and profitability.[10]

In some sense it would seem that Castel is talking about nothing more profound than the shift in the focus of public health intervention from infectious disease control to the management of 'lifestyle' related illnesses such as accidents, cancer and cardiovascular/circulatory system diseases: a shift resulting from the successful near-eradication of most high-mortality infectious diseases in Western nations. Or that the only difference between preventive instruments derived from a 'logic of dangerousness' and those derived from a 'logic of risk' is merely the difference between those intervening on an individual level and those intervening at the level of population: a difference notoriously related to the shift away from 'social concerns' during the 'age of bacteriology', and then back towards them under the aegis of the 'new public health'.[11] Both these readings are accurate to some extent, but they omit the subtler questions that inform the social study of risk as invited by Castel and others. We need to know not just that 'risk' equates roughly with 'population', but what social technologies have been used to govern it, and what political rationalities, from the welfare state to neo-liberalism, have deployed it.

Some years back, I argued there that the infection control strategies of the late nineteenth and early twentieth centuries were centered around concepts of cleanliness that lay at the heart of 'sanitary' public health, and were focused on locating and neutralizing all instances of dirt, in fact, every single bacterium.[12] That is, the logic of dangerousness did indeed underlie the strategies of notification, quarantine, isolation, disinfection and carrier control that were chiefly utilized at that time to control major killers such as diphtheria and typhoid fever. Pursuing the goal of eliminating all dangers did not seem impossible, merely gargantuan. In the first three decades of the twentieth century, excited by the instrumental prospects that were raised by new research in bacteriology, public health authorities harbored dreams of total hygienic containment, as when Park dreamed of swabbing the throats of all children in New York City in order to locate and control all diphtheria carriers, and thus eradicate the disease entirely.[13]

But as Castel suggested, the logic of dangerousness failed to achieve these goals. It proved impossible to reliably test enough people often enough to diagnose all the carriers, and even had this been accomplished, it proved impossible, short of killing them, to effectively neutralize the danger they posed. As a matter of fact, killing all living carriers of a danger *has* indeed been an instrument used, sometimes with success, among animal populations (for example, current extermination of poultry to control avian influenza strains). The same logic was also used horrifically and with no success by the Nazis, who attempted to eliminate all threats to what they saw as the integrity of their population by murdering all individuals who carried the danger of racial 'unfitness' in their bodies.[14]

The failure of these strategies led to what I argue were the first 'risk based' public health instruments, immunization and pasteurization. I term these 'risk based' because they were explicitly introduced *in order* to make the dangerousness or otherwise of individuals irrelevant. But they differed from the late twentieth-century calculative rationalities of risk that Castel talks about by their goal (to reduce risk to zero), the political dream that required them (the elimination of disease) and the political rationality that deployed them (the beginnings of welfarism through public insurance).

The shift from dangerousness to risk was not clear-cut historically. Moreover, in some cases a given preventive instrument may embody something of both logics, or be employed sometimes on the basis of one, and at others on the basis of the other. For example, sanitary public health in its mid nineteenth-century British heyday did aim at preventing the threats of alcoholism, crime and disease by changing the living conditions of the groups, not the individuals, that generated them. Immunization is likewise administered both on the basis of individual-level protection and population-level infection control. Though Castel claims the late twentieth century as a time when the notion of risk became 'autonomous' from that of danger, I suggest that in reality interventions have been designed at both individual and population levels, through both danger-destruction and calculation. The real key is where the emphasis has lain, and what difference that has made to public health practices, in different medico-political contexts.

I have continued to be interested in why and how the language and concepts of 'risk' have been employed at certain times in the history of public health, and with certain goals in mind. Especially interesting are the occasions in which scientists and public health authorities have grappled with the problem of identifying dangerous individuals who bear no signs of their dangerousness, and (not always the same) those in which authorities commit vast bureaucratic information–management and intervention resources to attempt to identify and confine all such individuals. These were both particular features of the outbreak of SARS in Toronto.

What were the logics of risk and danger in operation during the course of the outbreak of SARS in Toronto? What kinds of anxieties shaped authorities' responses to the event? What kinds of practices were mobilized for its containment, and what assumptions and expectations were at work during the course of the outbreak and its immediate aftermath?

SARS: an epidemic

To public health officials, SARS represented the return of an experience that most in the West had consigned to a past beyond memory: a swiftly spreading outbreak of an epidemic disease with a significant mortality rate, and no vaccine or specific cure. Canadians had not experienced an acute epidemic crisis since polio in the mid twentieth century. SARS' method of transmission was unknown, but its threat was all too easily imagined, mentally filtered through images of makeshift Spanish Influenza hospitals, crowded diphtheria wards, hospital rooms full of iron lungs.

In this section I retell the story of SARS in Toronto as it has been told in the major public inquiries into the crisis, supplemented by oral and published retrospectives from public health professionals.[15] I show that the response to SARS was built precisely around the particular risk posed by the dangerous but unmarked individual, feared because of how fast and through what momentary coincidences the disease seemed to spread. This response, utilizing the instruments of segregation and hygiene – quarantine, isolation, exclusion, disinfection – was mobilized as a result of the limitations of global information management instruments and in place of local epidemiological calculations, that is, across and above risk-based instrumentalities. The necessity of this response was a matter of debate in the inquiries, as stakeholders reflected on the consequences of the containment measures used. Where should the lines have been drawn?

The story begins with the failure of risk-based public health – the failure of surveillance. Information management is of course a central preventive instrument in risk-based health policy, since it is through systematic predetection and designation of at-risk populations that authorities can anticipate the irruption of danger. In the case of SARS, this should have operated within China, coordinated by the WHO, and within and across the rest of the world by monitoring specific at-risk populations, such as travelers to the affected area. But it didn't. For reasons vested in logistical limitations and human error, the information did not make it to appropriate managers in time. Canada's much admired Global Public Health Intelligence Network (GPHIN), also discussed in this volume, received a Chinese-language report of an influenza outbreak in mainland China in November, 2003. But the report was never translated. On 14 February 2004, the WHO reported in its weekly newsletter the occurrence of an outbreak of an unusual acute respiratory illness, thought by the Chinese authorities to be atypical pneumonia, in Guangdong province. On 21 February ProMED-Mail, an internet reporting system, noted that only two tissue samples from deceased patients in Guangdong had shown evidence of the pneumonia bacterium, and that the illness might not be pneumonia at all. These passed without comment, in part because the Chinese government suppressed information about the extent of the outbreak at this stage; Health Canada (the federal-level department of health) noted the outbreak only because of concern that it might presage a new and virulent form of avian influenza, the virus that has been concerning health officials for some years.[16]

While information was limited, a single moment of coincidental contact allowed the disease to spread. A doctor who had treated patients with atypical pneumonia in Guangdong then traveled to Hong Kong to attend a wedding, staying in the Metropole Hotel, where he became unwell and infected at least 12 other people. One was an elderly Toronto woman who had shared an elevator with him. (A similarly brief encounter in a hotel led to SARS' spread to Hanoi, where a patient was treated by the man who recognized the disease as novel and informed the WHO on 28 February, and died of it on 29 March: Dr Carlo Urbani). She became ill 2 days after her return to Canada, and died at home on 5 March. Her physician, noting nothing unusual in a very elderly woman's succumbing to a respiratory ailment, listed 'heart failure' as the cause of death. On 7 March, her 44-year-old son, Mr K., arrived in the emergency waiting-room of a major Toronto hospital, Scarborough Grace, with high fever and difficulty breathing. He was not identified as dangerous – for instance, he had no recent history of travel. He shared the open observation ward of a busy emergency department for 18 to 20 hours while awaiting admission and was later admitted to intensive care, where he required intubation. Before he was isolated (on suspicion of having tuberculosis), he received oxygen and vaporized medications, which are potentially capable of transforming infectious droplets into an infectious aerosol, and many patients, staff and visitors were exposed.

As the story of SARS in Toronto progresses, the emphasis remains on its contingency, eluding control because it was transmitted by people and objects not suspected to be dangerous. On 13 March, the WHO issued a global alert about the mysterious illness (characterized as primarily affecting health-care workers in Hanoi and Hong Kong, where it has spread to several hospitals), but many Canadian health practitioners remained unaware of the alert,[17] and in any case, by then it was too late. Mr K. died and his family, now ill, were infecting others, despite their fast isolation in negative pressure rooms in various Toronto hospitals. The story emphasizes the disease's ability to infect unpredictably and indiscriminately, in contrast to those many illnesses that mostly affect nicely demarcated at-risk groups such as the homeless, the immuno-compromised, or health staff. The microbe invaded across existing barriers, as apparently 'safe' equipment turned out to be dangerous: 'the physician who intubated Mr. P in the ICU wore a mask, eye protection, gown, and gloves while performing the procedure, but he developed SARS'.[18] The committee of senior physicians and medical officers for health that had responsibility for containing the epidemic, already fearful that they were witnessing 'another 1919', were literally terrified by the 30 March outbreak of 324 SARS cases amongst residents of vertically-linked apartments in a Hong Kong complex.[19] The committee's worst fears appeared to be confirmed on 12 April when a cluster of cases occurred in a Toronto Catholic sect.

Since danger seemed, potentially, everywhere and catastrophe was genuinely feared, the advisory committee recommended containment measures on an enormous scale. First, basic hygiene – hand washing and handkerchief use – was emphasized as the primary form of prevention to be undertaken by every Canadian. Secondly, 'social distance' measures were implemented. Schools closed if there was even the remotest chance of contamination. For example, one in Scarborough closed after a nurse's child exhibited symptoms. Excursions and events were canceled. At Easter, the Ontario Department of Health sought and received the cooperation of religious leaders, requesting they place communion wafers in hands rather than in mouths, refrain from using a common cup, hold confessions outside booths, and have parishioners exchange smiles rather than kisses or handshakes.

Thirdly, quarantine was used as a primary containing mechanism, and was imposed on 23,000 people in Toronto.[20] Anyone who had entered the affected hospitals after 16 March was asked to adhere to a 10-day home quarantine. Attendees of funerals were sent into quarantine; when an employee of a large information technology firm broke quarantine and returned to work with respiratory symptoms, 200 fellow employees were sent into home quarantine; when screening picked up a fever in a nurse caring for SARS patients, all the passengers who shared her train carriage that day were identified for possible quarantine. There were two levels of quarantine: 10-day home quarantine, and 10 or more day 'work' quarantine, where those affected could move between their workplace and their home, but go nowhere else. In addition, many of those in quarantine self-imposed extra barriers between themselves and their family, sleeping in basements and preparing and eating food alone.[21]

In point of fact, of course, the epidemic was almost wholly confined to health care settings, and those primarily affected were exposed and vulnerable patients and health care workers. A majority of containment measures therefore focused on these venues. Several family doctors were infected by their patients, leading to (highly unsatisfactory to all parties) discussions as to what protective clothing family doctors should use, and how it should be distributed to them.[22] Hospitals across Ontario – well outside the affected area – ceased undertaking non-emergency procedures, and closed their doors to visitors. Their borders were patrolled, sometimes literally by police, but at all times by staff who were deputed to temperature-screen all those who entered their doors and to require handwashing in alcohol-based antiseptic. Within, hygiene protocols received minute scrutiny. Standard protective clothing protocols quickly became double gown and glove. Health care workers worried anyway that they might contaminate themselves or others as they removed this clothing, or by other minor and inadvertent breaches of protocol. Masks became the focus of a lengthy and sometimes acrimonious debate over whether and to what degree fit-testing was required to make masks protective rather than an actual transmitter of infection: whether or not only the N95 mask (which filters out 95 per cent of all particles larger than 1 micron) was effective in preventing transmission, and how to manage the logistics of distribution and fitting.[23]

These containment measures had enormous negative social and economic impacts. First, the new protocols required in health care settings had tragic by-products. More people died from their inability to access full care during this period than died from SARS.[24] Family members prevented from visiting patients all over Ontario suffered anguish. Hospitals felt unable to implement all the Health Department directives, such as isolating all inpatients with fever or respiratory symptoms. Second, tracking those in quarantine was an enormous undertaking for public health officials. The WHO travel advisory against Toronto, which was issued in late April as the first outbreak was terminating, sparked off such a large number of self-reported 'possibles' that the masses of data helped obscure the case of the medical student who caused the second outbreak in May.[25] Third, the saturation media coverage, especially the 'SARS soap' – the daily 2 pm press briefing – as well as the disruptions wrought by school closures and Easter, contributed to a sense of crisis. Quarantine was immensely stressful, leading to anxiety attacks, nightmares and raised blood pressure in many subjects.[26] Finally quarantine also adversely affected already overstrained health resources. Four members of the advisory committee were quarantined when one of their members fell ill with SARS. Hospitals quickly lost staff with any experience to either illness or quarantine.[27]

The economic impacts of SARS were also drastic, especially after the WHO issued its travel advisory against Toronto. Airlines were nearly bankrupted.[28] As Strange details in her chapter, tourism to Ontario was decimated: one member of the SARS scientific advisory committee recalled staying at one of Toronto's major hotels, entirely vacant, and eating alone in its cavernous dining room.[29] Chinatown was emptied of shoppers; the entire hospitality industry, with flow-on effects to the arts and the provincial economy, significantly contracted. Quarantined workers and their industries lost income. Government spending on the crisis put a strain on the next budget. Such economic losses were in part the result of emotionally-damaging stigmatization effects. Nurses and their families felt socially isolated by nervous workmates and were refused service by frightened shopkeepers or taxi drivers. And many Torontonians identifying as Asian or of Asian descent complaining of feeling stigmatized.[30]

Fortunately SARS did not turn out to be the terrifying new plague. It was not in fact highly contagious, nor especially virulent in healthy people aged under 65. In the end, about 250 people were infected in Canada, and 44 died, which was and is a very small number in comparison with (for example) preventable deaths from tobacco smoking or road accidents. To some it certainly seemed a disproportionately small number of deaths in comparison with the disruption it caused, and various incompetencies, errors of judgments and over-reactions to risk were singled out as an explanation for what went 'wrong'.[31] In the analysis of SARS containment measures below I suggest instead that some of the impacts resulted from the sudden overwhelming commitment to old infection control measures – those based on the logic of dangerousness – as a result of acute fears that made more calculative methods seem unpalatable. These impacts were attendant on the enormous and often useless efforts to patrol borders and to sanitize boundaries and points of interaction.

Borders, risk and danger

SARS was frightening to many of those who had the responsibility of dealing with it.[32] In part it represented the threat of 'new and re-emerging infectious diseases' (commonly shortened to the American form of 'emerging infectious diseases or EID),[33] that has much exercised public health experts in Western nations over the past 15 years. In other words, SARS was frightening because there was a pre-existing conceptual and cultural space that made it so.

One key feature of current anxiety with EIDs is encapsulated in the reiterated phrase 'disease has no borders'.[34] This phrase was uttered many times in relation to SARS. Yet although 'borders are meaningless to a microbe', the first response was to require microbes to respect them anyway. In Western nations, the first move in combating EIDs has been to double the guard: to put in place more extensive and rigorous quarantine and border screening regimens, to examine prospective immigrants and exclude those believed to harbor illness, to identify, cordon off and patrol dangerous places. Why was this the case? In part because of the empirical requirements of infectious disease control, certainly, but in no small measure because of the return to a stance requiring the elimination, not merely the management, of the frightening danger of an EID.

Borders are a requirement of the logic of dangerousness. They were key to the more complicated public health strategies of the nineteenth and early twentieth centuries, which sought to define and separate the clean from the dirty, the immune from the susceptible, the infectious from the benign – categories that applied to people, objects, and spaces.[35] Bashford has explained how the boundaries of government have manifested as and through what she calls 'lines of hygiene', a designation that allows us conceptually to link the various practices of detention, segregation, isolation, quarantine, seclusion and bodily purification and management that have been the primary instruments of disease control for most of the past two centuries.[36] In public health policies largely built around the logic of controlling danger, boundary crossings were identified as sites requiring special policing and control.[37]

One manifestation of this in the SARS experience was the identification of air travelers as a special group in need of intense surveillance. In Canada, all passengers from SARS-affected countries were issued with a yellow leaflet, and outgoing passengers from Toronto with a red leaflet, containing a questionnaire of symptoms and recent activities. (These leaflets were available at some land borders also). Those who answered 'yes' to any question were examined by a nurse. By July 1 million people had received the leaflets and 3,000 had been examined. In addition, 800,000 people had been thermally scanned for raised temperatures and 200 had been examined further. None of these people were sent to seek further medical treatment. As it turned out, between March and May only five people infected with SARS entered the country, and none were symptomatic whilst flying or in airports.[38] In fact screening travelers nationally and internationally – millions were screened worldwide – turned out to be a colossally useless procedure. Nevertheless air travel continued to be a strong focus for concern, as demonstrated by the controversial travel advisory issued against Toronto by the WHO (which required thermal scanning be implemented at all international airports), backed by the US Centers for Disease Control (CDC).[39]

Similarly, the Canadian response to SARS was instrumentally about the almost impossible task of monitoring a nearly infinite multitude of potentially dangerous objects and people. The experts who sat on the SARS Scientific Advisory Committee, terrified that they would become 'the people who failed to stop SARS',[40] were faced with precisely the dilemma that Foucault described in relation to the psychopath: how to identify and neutralize the dangerous before they showed any sign of their danger? Which among the heterogenous group of nurses, gardeners, cooks, and lawyers would get sick? Which gloves and masks and tubes would transmit the virus instead of protecting the sufferer? Which momentary and coincidental proximities would spread disease? This multiplication of fears was directly responsible for members of the SARS Scientific Advisory Committee in Toronto feeling compelled to take up any and all instruments that might confine the danger – hence the hospital closures and heavy reliance on quarantine.

One person did query the logic of this response. Dr Richard Schabas, a former Chief Medical Officer of Health in Ontario and Chief of Staff of a SARS-affected hospital during the outbreak, questioned the use and extent of quarantine and contact tracing on the basis that epidemiologically, the disease had *never* looked like a highly contagious epidemic, even in Guangdong. He argued that quarantine was at best a poorly understood and at worst largely useless infection control instrument. Its imposition was arbitrary,[41] its length, and its subjects, being determined by social rather than scientific factors.[42] (Different SARS-affected countries used vastly different quarantine regimens; those that quarantined fewer contacts than the ten per suspected case in Canada did not suffer increased infection rates.) Further, tracing the epidemic curve – a primary epidemiological tool – showed the outbreak had peaked and was declining by the end of *March*.[43] Even Schabas could not bear to risk passing on the infection, and voluntarily entered quarantine for the duration of his holiday in France when he became aware that a SARS patient had entered his hospital around the time he left Canada. Nonetheless, Schabas was publicly skeptical of what he saw as a kind of disaster mindset in the scientific advisory committee, wondering if there was a need to 'worry about the hundred year flood every time it starts to rain'.[44] Schabas felt fears of worst-case scenarios and of unlikely contingencies were being chosen over the more minimal rationales suggested by risk assessment.

In summary, the scientific advisory committee tried to eliminate all potential sources of danger regardless of the consequences, rather than accepting the possible casualties of an approach based on calculating probabilities. And the committee felt justified when the second outbreak in

May, SARS II, occurred after *two full incubation periods* had passed with no new cases. This was transmitted by a medical student, exactly the kind of tiny probability/high consequence event they were worried about.[45] Faced with the acute danger represented by a new infectious disease, public health experts returned to the fantasy of total hygienic containment that has periodically animated public health instrumentalities. Partly as a result of SARS' novelty but partly also out of sheer anxiety, highly conservative danger-containing measures were taken based on very little evidence; they were agonized over, but the key value – preventing further spread of SARS – was never questioned in cost/benefit analyses. Because they were forced to take action at the individual and not the population level this fantasy required the expenditure of almost unthinkable resources in time and money. Economic impacts were incurred as a direct result of the use of travel screening, quarantine and social distance measures, and stigmatization effects can be understood as a logical corollary of the search for the dangerous. And in the end, this was largely accepted as the shape of things to come.

The 'new normal'? An insecure conclusion

Since SARS it seems we are living in a 'new normal'. The term has had wide circulation in the media and in specialist circles after both the attacks on the World Trade Center on 11 September 2001,[46] and after SARS. It is impossible to dissociate SARS from 9/11 – both are consistently used, often together, as emblems for the novel, catastrophic threats that North America and the 'west' are considered to be facing. The words 'new normal' speak to the new discourses of (in)security and the methods used to combat danger in an era defined by these two events. In both cases the uncertainty of the threat posed by individuals who show no sign of their dangerousness and who may do harm at any time has been represented as especially dreadful, and in both cases has led to minute and scrupulous border patrol, the notion that security is vested in and reinforces the nation. This is, in other words, the attempt to confine in order to prevent.

In Toronto, the 'new normal' refers particularly to health issues. In its broadest terms, it refers to the EIDs, a threat represented as equivalent to terrorism. Insofar as SARS represented this threat, border patrols and detaining the dangerous were central instruments. During and after the outbreak, however, the term 'the new normal' had more specific meanings, referring to (1) new directives for infection control within hospitals,[47] and (2) a call for all members of the public to return to the principles and practices of basic hygiene, which were considered to have lapsed under conditions of false security.[48] (These were emphasized in the emergency preparedness plans for businesses, also).[49] The new directives aimed specifically for hospital infection control had been issued and updated by the

Ontario Ministry for Health during the course of the outbreak, and centered on more scrupulous sterilization procedures, the wearing of new multiple layers of protective clothing and especially fit-tested N95 masks, and the isolation in negative pressure rooms of all patients with fever or respiratory illness. Although they were devised in part to respond to the fears of frontline health care workers, these directives were initially found unworkable in existing hospital conditions and with existing resources, and since then have been revised with input from public health officials, doctors and nurses.[50] This 'new normal' of increased vigilance over a range of tiny details that formerly went unscrutinized – a hospital version of border security – continues to be operational.[51]

In both contexts, all sides agree that security comes at the price of a vanishing ease and innocence. Practitioners wonder if they should greet all their patients swathed in greens, their faces sweating and itchy, their expressions masked: 'The bare-faced examiner of coughing children should be considered an image of the past, seen only in Norman Rockwell paintings of a simpler time', wrote one doctor. 'Believe me it's not ideal, and if I had the choice I wouldn't be doing it. But I think we have an obligation to protect ourselves and our patients given what we know', commented another.[52]

However some physicians acknowledged a calculative aspect to their decision-making about infection control:

> I'll have to try to figure out what to do when I get to work. It looks like we're going to have the sign-in sheets for everyone, and I have masks and alcohol gel at the front door for patients. I still don't know if I should be wearing a mask all the time – they're very uncomfortable. I think I'll have a look at what the patient has written on the sheet before entering the room, and put on a mask if I'm worried. Not very high tech, but I hope it works. I guess this is the new normal I've been hearing about.[53]

In fact, though I have dwelt on the reasons and ramifications of a return to the stance of danger-control in the response to SARS, in the 'new normal' instrumentalities based on the logic of risk co-exist and interact with those based on the logic of controlling danger. As in the case above, this is so if for no other reason than that both stances have their failures, making each reliant on the other. At the mundane level, decisions about whether or not to cancel concerts, quarantine the contact of a contact, screen at this airport but not that land border, or prevent a daughter from visiting her dying father in a hospital 8 hours' drive from Toronto, are calculated as risks, but only operationalized through spatial instruments of confining and segregating. This has the effect of reproducing borders and boundaries between the safe and the dangerous, the clean and unclean, resulting inter alia in stigmatization effects.[54]

The point becomes clearer if we look at the global context in which SARS occurred. Certainly nations responded by defending their borders from the infected. But at the same time the response to SARS saw unprecedented surveillance and intervention by WHO in the internal response mechanisms of sovereign nations. As other scholars in this book demonstrate, this has been seen as representative of a global shift away from the 'Westphalian' model of international public health, in which explicitly autonomous nation-states interact to control the spread of disease from one to another, to 'global health governance', in which it is legitimate for given authorities, such as the WHO, to use non-government sources of information and to intervene within a state to prevent and contain disease.[55] Global health governance combines instrumentalities of risk and danger by necessity. International organizations operate calculative information surveillance so as to predict the advent of danger, as the WHO is currently doing with respect to pandemic influenza. But they also utilize containment (danger elimination) instruments to confine danger, where it threatens in the shape of a highly infectious disease. And despite the humanitarian potential of the WHO, these processes do not occur in an equal or altruistic world. The new 'hyperpower', the US, operates a kind of global factor management, trading off economic benefits with health impacts. But also, during and after SARS, the US emphasized border controls, quarantine and exclusion to eliminate danger. This is what Nick King refers to as the 'EID worldview', a US-centric approach to biosecurity in which national borders are reproduced, as they were during SARS, yet deterritorialization is used as a resource for extending US security and hegemony.[56]

For people who live in Western nations, the 'new normal' means not global interconnectedness but a new way of defining citizenship. The truly healthy subject–citizen of the 'new normal' is a person who not only exercises a disciplined and responsible autonomy, managing and minimizing the risks to their health from their lifestyle, obedient to the calculative rationality of risk-based managerial public health.[57] This person is also constantly vigilant for catastrophe, and immediately obedient to the imposition of instruments used to identify and eliminate the dangerous. The American Red Cross, for example, has produced a new initiative called 'Together We Prepare', which 'champions Individual Preparedness'. The initiative 'empowers people to prepare themselves, their homes, schools, businesses, and neighbourhoods for the unexpected'.[58]

The 'new normal' is a state of accepted economic and personal insecurity driven by the demands of the global marketplace. Governments have little heeded the calls for risk reduction through a 'renewal of public health' and the redress of social inequalities. Instead they are developing emergency preparedness plans to confront new epidemics whenever they should occur. The dilemmas remain: the impossibility of locating and neutralizing all dangerous individuals and objects in real time, the inappropriateness of making judgments based on risk calculations where mistakes cost lives in a crisis. In resolving this, the lines of hygiene that divide the world are likely to be redrawn many times.

Notes

This research was funded by the National Health and Medical Research Council of Australia Sidney Sax Post-doctoral Fellowship in Public Health.

1 H. Branswell, 'Study suggests widespread quarantine not needed for SARS containment', *Canadian Press NewsWire* (30 October 2003).
2 H. Branswell, 'U.S. public health chief says emergence of SARS, avian flu the "new normal"', *Canadian Press NewsWire* (25 February 2004).
3 R. Castel, 'From Dangerousness to Risk' in G. Burchell, C. Gordon and P. Miller (eds) *The Foucault Effect: Studies in Governmentality* (London: Harvester Wheatsheaf, 1991), pp. 281–99.
4 S. Page, 'The Ottawa Citizen's Shelley Page writes from southern China about adopting a baby girl against a backdrop of the growing SARS crisis', *CanWest News* (25 April 2003).
5 Castel, 'From Dangerousness to Risk'.
6 M. Lewis, *Thorns on the Rose: the history of sexually transmitted diseases in Australia in international perspective* (Canberra: Australian Government Publishing Service, 1998).
7 V. Berridge, 'Science and Policy: The case of postwar British smoking policy' in S. Lock, L. Reynolds and E. Tansey (eds), *Ashes to Ashes: The history of smoking and health* (Amsterdam: Clio Medica, 1998).
8 R. Bunton and A. Petersen, *Foucault, Health and Medicine* (London and New York: Routledge, 1997); A. Petersen and D. Lupton, *The New Public Health: Health and self in the age of risk* (Sydney: Allen & Unwin, 1996).
9 Petersen and Lupton, *The New Public Health.*
10 Castel, 'From Dangerousness to Risk'.
11 Petersen and Lupton, *The New Public Health.*
12 C. Hooker, 'Sanitary Failure and Risk: Pasteurisation, immunisation and the logics of prevention', in A. Bashford and C. Hooker (eds), *Contagion: Historical and Cultural Studies* (London: Routledge, 2001), pp. 129–52.
13 E. Hammonds, *Childhood's Deadly Scourge: The Campaign to Control Diphtheria in New York City, 1880–1930* (Baltimore: Johns Hopkins University Press, 1999).
14 R. Proctor, *Racial Hygiene: Medicine under the Nazis* (Cambridge, Mass: Harvard University Press, 1988); Castel, 'From Dangerousness to Risk'.
15 My primary sources are comments made by health professionals and policy makers in the news media, the two Reports of the commissions of inquiry into SARS that were called by the national and provincial governments (respectively the Naylor and Campbell Reports), reflections on SARS published by health professionals in their own journals, and interviews I conducted with a majority of members of the SARS scientific advisory committee, who shall not be individually named. A. Campbell, *SARS and Public Health in Ontario* (Toronto: Ontario Ministry of Health and Aged Care, 2004) http://www.sarscommission.ca (accessed 23 November 2005); D. Naylor, *Learning From SARS: Renewal of Public Health in Canada* (Ottawa: The National Advisory Committee On SARS and Public Health, Health Canada, 2004) www.phac-aspc.gc.ca/publicat/sars-sras/naylor (accessed 23 November 2005).

16 Naylor, *Learning From SARS.*
17 *Ibid.*
18 *Ibid.*
19 Interview data.
20 Naylor, *Learning From SARS*; T. Svoboda et al., 'Public Health Measures to Control the Spread of the Severe Acute Respiratory Syndrome during the Outbreak in Toronto', *New England Journal of Medicine*, 350 (2004): 2352–61.
21 L. Hawryluck, W. Gold, S. Robinson, S. Pogorski, S. Galea and R. Styra, 'SARS Control and Psychological Effects of Quarantine, Toronto, Canada', *Emerging Infectious Diseases*, 10 (2004): 1206–12.
22 Naylor, *Learning From SARS.*
23 *Ibid.* S. Nicholls, 'A new normal? Across the nation, doctors are grappling with exactly what regular practice will be in a world with SARS', *Medical Post*, 39, no. 23 (10 June 2003); M. Schull and D. Redelmeier, 'Infection control for the disinterested', *Canadian Medical Association Journal*, 169 (2003): 122–3.
24 B. Sibbald, 'SARS' other toll', *Canadian Medical Association Journal*, 168 (2003): 1697.
25 Svoboda et al., 'Public Health Measures'.
26 Hawryluck et al., 'SARS Control and Psychological Effects of Quarantine'.
27 Naylor, *Learning From SARS.*
28 'War and SARS continued to take their toll on travel to and from Canada in April, with Ontario particularly hard hit, Statistics Canada says', *Canadian Press NewsWire* (19 July 2003).
29 Interview data.
30 C. Leung, *Yellow Peril Revisited: Impact of SARS on the Chinese and Southeast Asian Canadian Communities* (Toronto: Chinese Canadian National Council, 2004).
31 H. Skinner, 'The Fog of SARS', paper presented at *Learning From SARS: Challenges to Public Health*, University of Toronto, 2003.
32 Interview data.
33 N. King, 'Security, Disease, Commerce: Ideologies of Postcolonial Public Health', *Social Studies of Science*, 32 (2002): 763–89.
34 L. Kumove, 'Raiding poor countries', *Community Action*, 18 August 2003; S. Rutton, 'Feds wants more control during health crisis', *CanWest News* (9 June 2003).
35 A. Bashford, *Imperial Hygiene: A critical history of colonialism, nationalism and public health* (London: Palgrave, 2004); A. Bashford, *Purity and Pollution: Gender, embodiment, and Victorian medicine* (London: Macmillan, 1998).
36 Bashford, *Imperial Hygiene.*
37 W. Anderson, *The Cultivation of Whiteness: Science, health and racial destiny in Australia* (New York: Basic Books, 2003); D. Armstrong, 'Public Health Spaces and the Fabrication of Identity', *Sociology*, 27 (1993): 393–403.
38 R. St John, A. King, D. de Jong, M. Bodie-Collins, S. Squires and T. Tam, 'Border Screening for SARS' *Emerging Infectious Diseases*, 11, no. 1 (2005) http://www.cdc.gov/ncidod/EID/vol11no01/pdfs/04–0835.pdf (accessed 23 November 2005).
39 C. Laino, 'Travellers getting caught in SARS net: tell patients to keep vaccinations up to date and practise good hygiene to avoid common respiratory illnesses that may be mistaken for SARS', *Medical Post*, 39 (2003): 15; Naylor, *Learning From SARS.*
40 Interview data.

41 S. Page, 'The Ottawa Citizen's Shelley Page writes from southern China about adopting a baby girl against a backdrop of the growing SARS crisis', *CanWest News* (25 April 2003).

42 Branswell, 'Study suggests widespread quarantine not needed for SARS containment', *Canadian Press NewsWire* (30 October 2003); S. Staples, 'New formula designed to predict how long patients need to be quarantined', *CanWest News* (29 June 2004).

43 R. Schabas, 'Prudence, not Panic', *Canadian Medical Association Journal*, 168 (2003): 1432–4.

44 Interview with Dr Richard Schabas, 14 July 2004.

45 D. Hawaleishka, 'A new SARS bombshell', *Maclean's* (2 June 2003).

46 R. Simon, J. Howe, K. Reese, L. Huriash and N. Neuser, 'The New Normal; A nation already on high alert braces for the possibility of new attacks', *U.S. News & World Report*, 131, no. 20 (2001): 14–16.

47 'Rouge Valley hospitals ready for "new normal"', *Canada NewsWire* (22 May 2003).

48 A. Stelmakowich, 'Response to SARS: Emergence of the new normal', *OH & S Canada*, 19, no. 4 (2003): 48–9.

49 B. Orr, 'SARS outbreak teaches valuable lessons on a new "normal" state for HR management', *Canadian HR Reporter*, 16, no. 11 (2003): 5.

50 G. Bonnell, 'SARS directives still ignore expertise of health-care workers, hearing told', *Canadian Press NewsWire* (30 September 2003); Nicholls, 'A new normal?'

51 'Ontario man monitored for respiratory ailment; probability of SARS "low"', *Canadian Press NewsWire* (10 May 2004).

52 S. Kirkey, 'Doctors told to wear face masks and goggles when examining coughing youngsters', *CanWest News* (10 December 2003).

53 M. Grieve, 'The new normal: a SARS diary', *Canadian Medical Association Journal*, 169 (2003): 1283–5.

54 N. Pidgeon, R. Kasperson and P. Slovic (eds) *The Social Amplification of Risk* (Cambridge: Cambridge University Press, 2003).

55 D. Fidler, *SARS, governance and the globalization of disease* (London: Palgrave, 2004).

56 King, 'Security, Disease, Commerce'.

57 Peterson and Lupton, *The New Public Health*; Bunton and Petersen, *Foucault, Health and Medicine*.

58 'American Red Cross President to deliver Keynote Address at 15th World Conference on Disaster Management in Toronto', *Canada NewsWire* (8 November 2004).

11
Biosecurity: Friend or Foe for Public Health Governance?

David P. Fidler

The annals of public health record many milestones in the evolution of this area of the governance of human affairs. The development of bacteriology in the late nineteenth and early twentieth centuries represents one of these transformational events that changed the theory and practice of public health forever. Other seminal changes include the emergence of international cooperation on public health in the late nineteenth and early twentieth centuries, the development of antibiotics and vaccines in the mid-twentieth century, and the convergence of public health and the human rights movement in the second half of the twentieth century. This chapter analyzes another transformational moment in public health as a governance activity – the emergence of biosecurity.

The rise of 'biosecurity' as a policy and governance issue in the last decade represents one of the most profound changes to affect public health in its long history. The shift in public health caused by biosecurity is still so recent that its importance and portents remain unsettled and ambiguous. This chapter's major purpose is to explain the significance of the emergence of biosecurity as a policy issue and to describe the new public health governance reality created by biosecurity's emergence. Biosecurity has changed public health forever; but the nature of this change remains, in many respects, to be determined by how states, international organizations, and non-state actors shape this transformation in the years ahead.

My analysis proceeds in four parts. First, I define 'biosecurity' to ensure that the reader understands the concept as used in this chapter. Second, I provide an historical overview of the relationship between national and international security and infectious diseases. This overview develops context that is necessary for grasping the policy revolution biosecurity represents. Third, the chapter describes the emergence of biosecurity over the past decade. Finally, I analyze the significant impact the rise of biosecurity has on public health governance nationally and globally. I draw examples from biosecurity policy in the United States where the pursuit of biosecurity is prominent and advanced. This analysis involves assessing whether the current trends in biosecurity policy help or hinder the task of public health governance in the early twenty-first century.

Defining 'biosecurity'

Although increasingly used in policy debates and academic discourse, 'biosecurity' does not have an agreed meaning. The diversity of definitions reveals the messiness that develops when new concepts emerge, but this diversity also reveals deeper issues at work in the evolution of the concept. The most common definition associates biosecurity with efforts to protect states from attacks using biological weapons (bioweapons), with emphasis on securing a nation against biological terrorism (bioterrorism). In the United States, for example, biosecurity forms part of the larger strategy of strengthening homeland security: preventing terrorists' attacks on the United States, protecting US citizens and critical infrastructure from terrorist attacks, and preparing to respond effectively to terrorists' attacks that might occur.[1]

A broader definition of biosecurity incorporates security threats not only from bioweapons but also from naturally occurring infectious diseases. This more expansive interpretation captures arguments made in the last decade that severe epidemics of infectious diseases can constitute threats to national and international security. Both the Clinton and Bush administrations advanced the notion that naturally occurring infectious diseases can threaten national and international security.[2] The World Health Organization (WHO) has also connected threats posed by infectious diseases with the concept of security through its strategy to achieve 'global health security'.[3]

This chapter uses the broad definition of biosecurity because the profound transformation that public health has experienced in the past 10 years has touched virtually every aspect of public health's involvement with infectious diseases. The emergence of biosecurity simultaneously affected concerns about bioweapons and naturally occurring infectious diseases, making it artificial to try to separate the two areas when defining biosecurity. Using the broad definition of biosecurity does not mean, however, that biosecurity policies currently unfolding give equal weight to bioweapons and naturally occurring infectious diseases. As explored below, one of the 'fault lines' in biosecurity policy involves controversies about whether attention paid to bioweapons discounts the threat presented by naturally occurring pathogenic threats.

The historical relationship between security and infectious diseases

My claim that biosecurity constitutes a transformational moment for public health governance does not mean that security concerns about infectious diseases have never arisen before. This section examines the historical relationship between security and infectious diseases in order to provide the background necessary for the reader to understand the transformation biosecurity triggered for public health. Traditional concepts of security dominate this historical relationship, which means the relationship focused on states' concerns with how infectious diseases connected to fears about external military threats to state power and capabilities in the international system. Such traditional framing of the security concept meant that security concerns with infectious diseases infrequently connected with the theory and practice of public health within states.

Experts have often observed that military forces during wartime historically lost more troops to infectious diseases than to combat operations. The pathogenic threat to preparedness and power caused national armed forces to focus on military health. In the late nineteenth century, militaries implemented sanitary reforms in order to reduce the toll infectious diseases imposed on fighting forces. The Prussian military was particularly effective with its sanitary reforms: its forces suffered fewer deaths from infectious disease than battlefield deaths during the 1870–1871 Franco-Prussian War. This achievement was the first time in European military history that infectious diseases did not kill more soldiers than combat.[4]

Military interest in reducing infectious disease morbidity and mortality also appeared in connection with efforts by countries to maintain and expand their imperial territories or realms of hegemonic influence. The efforts made, for example, by the United States Army to address yellow fever connected to the Army's role in maintaining US military and economic hegemony in the Western hemisphere.[5] Likewise, interest in tropical diseases emerged as European countries deepened and broadened their empires in tropical regions around the world.[6]

Although military efforts against infectious diseases produced some benefits for civilian public health,[7] the objective of military health was to preserve the readiness and strength of the armed forces as one of the critical attributes of state power in the international system. The competition for power among states in international relations that drove military interest in infectious diseases sometimes influenced how states framed civilian public health approaches to the threat of disease importation from other countries. In the late nineteenth and early twentieth centuries, for example, Germany feared 'an epidemic spreading like wildfire from the east [that] could attack the nation's political, military, cultural, and economic superiority', producing a situation in which 'Germany's aspirations as a world power in commerce and international politics spurred on the build-up of its bacteriological defences'.[8]

The second context in which security and infectious diseases had a relationship concerns the potential use of bioweapons by states during war. Late nineteenth-century advances in the science of infectious diseases created the possibility that states might develop and use biological agents as weapons in times of war. Crude efforts to use infectious diseases as weapons populate the history of warfare prior to the twentieth century, but the scientific advances made in the late nineteenth and early twentieth centuries made conceivable more sophisticated weaponization of biological agents. States responded to this new threat to military forces and the conduct of war by prohibiting the first use of bioweapons in the Geneva Protocol of 1925.[9] The Geneva Protocol did not prohibit the development of bioweapons; nor did it ban the use of bioweapons against (1) countries not party to the Geneva Protocol; and (2) states parties to the Geneva Protocol that violated the Protocol's first-use prohibition. The Geneva Protocol put use of bioweapons firmly within the corpus of the laws of war, an indication that states believed that bioweapons presented a clear and present danger to their military power and preparedness.

Embedding bioweapons within the laws of war had, however, no implications for public health generally in the 1920s and 1930s. Concerns with the possibility of battlefield use of bioweapons, exacerbated by the development of long-range airpower, did not produce national security interest in the quality of the civilian public health infrastructure inside states. The issue of bioweapons' use by states remained within the traditional worlds of foreign policy and national security.

The third aspect of the historical relationship between security and infectious diseases involves arms control. In 1972, states, including the United States and the Soviet Union, adopted the Biological Weapons Convention (BWC), which banned the development, production, and stockpiling of biological and toxin weapons.[10] Countries joining the BWC agreed to end, or forgo, research into, and development of, bioweapons, an approach that terminated decades of efforts by states to build such capabilities.

Public health was not the catalyst for these events, although ending bioweapons programs would clearly benefit it. The decisive factor was the US government's unilateral determination in the late 1960s that bioweapons had little military utility and thus could be abandoned without threatening US national security interests.[11] This calculation flowed from traditional views of national security and military power and owed nothing to the development of public health in the decades following the Geneva Protocol. If public health factored at all into biological arms control, it was as a source of friction. International legal prohibitions on bioweapons development could not outlaw peaceful research on, and uses of, biological agents because such activities were necessary for public health purposes, namely scientific research on pathogens and applied research on, and development of, drugs and vaccines. The friction arose because drawing a clear line between peaceful scientific research on pathogens for public health purposes and research on offensive bioweapons capabilities has always been difficult.

The historical relationship between security and infectious diseases showed no signs that military and national security concerns with bioweapons involved public health or its governance. The development of public health during this historical period exhibits the same separation of public health from security concerns with bioweapons. For example, national and international public health systems operated without any reference or connection to national and international security concerns about bioweapons. States never applied the international legal regime for infectious disease control in ways supportive of national and international security concerns about bioweapons. Nor were national and international public health systems and resources designed or applied with the threat of bioweapons in mind.

One connection between public health and bioweapons appeared briefly after World War II in the United States because of fears of possible Soviet use of bioweapons against the United States. This fear helped stimulate development of US federal public health capabilities, such as the Epidemiological Intelligence Service.[12] This integration of public health and national security proved short-lived, however, as the two policy areas subsequently developed along different tracks with no interdependence.

The separate development of the policy spheres of security and public health was not irrational or illogical but reflected the nature of the threats posed by bioweapons, on the one hand, and naturally occurring infectious diseases, on the other. The prospect that a state would use bioweapons against the troops or civilian population of another state has always been remote. Japan used bioweapons against Chinese troops and civilians before and during World War II; but no state – whatever its ideology – subsequently joined Japan as a country that used bioweapons in armed conflict. Although many countries were involved in developing bioweapons after World War II, these development programs did not trigger serious fears that use of such weapons was imminent or even likely. Feeding into this perception was the realization that developing effective bioweapons was scientifically difficult, even for countries with sophisticated scientific talent and resources.

The US decision in the late 1960s to abandon unilaterally its offensive bioweapons program reinforced the separation of security and public health as policy endeavors. Although the Soviet Union accelerated its offensive bioweapons program after the United States terminated its efforts, the US move, combined with the BWC's adoption, took the biological-weapons edge off Cold War international politics. Thus, incentives for extensive and sustained public health preparations, especially for civilian populations, to address possible bioweapons use were minimal, if not non-existent.

In terms of naturally occurring infectious diseases, a number of factors explain why such diseases were decreasingly considered security threats in a military or geopolitical sense. First, most of the twentieth century witnessed progress by developed countries in reducing morbidity and mortality caused by infectious diseases. Such reductions widened the gap between public health and security as infectious diseases faded as a threat to public health in more affluent countries. Second, as noted earlier, security in the twentieth century was associated with traditional concerns about external military threats calculated through the geopolitical balance of power. The idea that naturally occurring infectious disease outbreaks could threaten national security, as traditionally defined, would have only made sense if outbreaks could adversely affect a country's military power and preparedness vis-à-vis a dangerous rival state. The only infectious disease outbreak in the first 80 years of the twentieth century that might have qualified as a national security threat in this sense was the 1918–1919 influenza pandemic, but this calamity did not produce efforts to broaden the concept of national security beyond its traditional military and geopolitical focus.

Third, public health activities at the international level underwent significant changes in the post-World War II period. From the mid-nineteenth century until World War II, the effort to ensure that national public health measures, such as quarantine, did not unnecessarily restrict flows of international trade and commerce of powerful trading nations dominated international health activities.[13] This balancing act was not concerned with improving health conditions inside developing countries or regions. After World War II, the theory and practice of international health changed dramatically.

This change can be sensed in the principles contained in the preamble of the WHO Constitution.[14] It stated, for example, that the enjoyment of the highest attainable standard of health is a fundamental human right. This principle shifted the focus of international health activities from the trade interests of the great powers to the health conditions affecting individuals. The WHO Constitution's preamble also contains a concept of security that bears little resemblance with the traditional definition of this notion in foreign policy and international relations. The preamble interprets security to include health: 'The health of all peoples is fundamental to the attainment of peace and security'.[15] Security in this perspective is not military security embedded within a balance of power but what is later in the century termed 'human security'.

Although the WHO Constitution contained a new concept of security that incorporated public health, the concept had little impact on how states and foreign policy makers conceived of security or public health. The Cold War struggle between the United States and the Soviet Union kept traditional interpretations of security dominant. In its activities, however, WHO pursued the vision set out in the preamble of its Constitution by turning its attention to improving health conditions in developing countries and regions, as illustrated by the 'Health for All' campaign launched in the late 1970s.[16] This direction in international health fed perceptions in foreign policy thinking that public health was 'mere humanitarianism', a form of 'low politics' peripheral to the real game of international politics. With the Cold War hardening acceptance of traditional concepts of security, and with WHO shifting international health away from the trade interests of powerful countries toward efforts aimed at helping those most in need, public health and security continued to go their separate ways as policy endeavors.

Policy worlds collide, biosecurity emerges

The policy worlds of security and public health collided after the Cold War because of developments in the areas of bioweapons and naturally occurring infectious diseases. The scientific, political, and moral restraints that made state use of bioweapons against another state during the Cold War unlikely did not exhibit the same strength in the post-Cold War era. Although still scientifically difficult, the challenges of developing bioweapons were receding with rapid advancements in microbiology and their global dissemination. The old political and moral constraints deterring states from using bioweapons might not hold if terrorists contemplated the malevolent use of microbes.

The growing scientific and political possibility of bioterrorism generated a new kind of security threat in the post-Cold War world. The policy responses to bioterrorism could not rely on deterrence but had to include preparing for the actual use of a biological weapon against civilian populations. In this context, the quality of national and international public health capabilities became a critical security concern for states and the international system. When national security communities realized the importance of public health to defense against bioterrorism, the gap between these policy worlds closed rapidly.

The world of public health changed not only with the rise of the bioterrorism threat but also with recognition of a global crisis in emerging and re-emerging infectious diseases. After decades of complacency, developed and developing states realized that a complex set of factors was driving the resurgence of infectious diseases.[17] The complacency that individuals and populations had developed concerning infectious diseases dissipated; and, in connection with the HIV/AIDS pandemic, states in the developing world faced microbial-related destruction of their populations, economies, development prospects, and military power and preparedness.

The resurgence of infectious disease encouraged security and public health experts to broaden the notion of security to include naturally occurring infectious diseases. This broadening occurred simultaneously with more general efforts to develop the concept of 'human security', which attempted to refocus security thinking on threats faced by people and individuals in their daily lives rather than on states, their military power, and the potential for war.[18] Expanding the concept of security to include infectious disease threats produced the policy need to strengthen public health infrastructure nationally and internationally.

The threat of bioterrorism propelled security thinking toward public health, and the resurgence of infectious diseases launched public health toward the realm of security. These changes gathered momentum from the mid-1990s on. The collision of the policy worlds of security and public health produces the new policy space called biosecurity. This, in turn, has affected how both security and public health are conceptualized and pursued. The threats posed by bioterrorism and naturally occurring infectious diseases make the quality of public health capabilities a central political and governance concern nationally and internationally. The new domain of biosecurity elevates public health's political status, melding together the former 'high politics' of bioweapons and 'low politics' of naturally occurring infectious diseases into a new type of politics without historical precedent.

The importance of biosecurity as a product of the collision of the policy worlds of security and public health can be illustrated by reference to the *Report of the Secretary-General's High-Level Panel on Threats, Challenges, and Change* issued in December 2004.[19] In this *Report*, the UN High-Level Panel noted the traditional focus of 'collective security' within the UN on the security of the state from a military perspective; but the Panel proceeded to develop the concept of 'comprehensive collective security'. It defined a threat to international security to involve '[a]ny event or process that leads to large-scale death or lessening of life chances and undermines States as the basic unit of the international system'.[20] The UN Secretary-General applauded the High-Level Panel's finding that '[w]e need to pay much closer attention to biological security'.[21]

Infectious diseases feature prominently in two threat clusters identified by the UN High-Level Panel – bioweapons and naturally occurring infectious diseases. The UN Secretary-General's own report on reform of the United Nations also stressed the importance of addressing naturally occurring infectious diseases and the threat posed by bioweapons.[22] These threat clusters are the core concerns of biosecurity. With respect to these threats, the UN High-Level Panel and the Secretary-General urged the strengthening of national and international public health capabilities. These recommendations underscore the importance of public health governance to the UN High-Level Panel's and Secretary-General's concept of comprehensive collective security. Although the comprehensive collective security concept encompasses more than biosecurity, the prominence of biosecurity threats, and the necessity of addressing them effectively, found in these reports stand as important evidence that a policy revolution has taken place, the nature of which we are still trying to comprehend.

Biosecurity and public health governance

The emergence of biosecurity as a policy issue has generated controversies about the most appropriate ways to respond to the threats of bioweapons and re-emerging infectious diseases. Some controversies involve different evaluations about whether the impact of biosecurity on the theory and practice of public health is positive or negative. Although consensus exists that biosecurity has transformed public health, agreement on the meaning of the transformation for public health has not emerged to date. Arguments stressing the positive impact of biosecurity's emergence note that the coming-together of the policy worlds of security and public health has brought public health political attention and financial resources it would never have received otherwise. Biosecurity has elevated public health from the margins of 'low politics' to a seat at the table of the 'high politics' of national security, foreign policy, and international relations.[23]

Negative reactions to biosecurity's rise include concerns that biosecurity has transformed public health's general mission to protect and promote human health into one fixated on defending against very low-probability bioterrorist attacks. Biosecurity mutates public health's traditional humanitarian ethos into a mentality that echoes conventional, narrow interpretations of security. Thus, tremendous amounts of political and economic capital are devoted to defending against bioterrorist attacks, the actual risk of which may not support such massive efforts. In short, biosecurity represents just another form of public health's subordination to traditional notions of national and international security.[24]

Understanding whether biosecurity represents 'friend or foe' for public health requires digging into the impact that biosecurity has had, or might have, on the governance of public health activities. This task leads analysis into considerations of how biosecurity affects key public health governance functions, such as surveillance. Analysis of these functions reveals two major fault lines in the emerging area of biosecurity: a fault line between (1) bioweapons and naturally occurring infectious diseases; and (2) a national and an international focus on biosecurity. The fault lines allow us to see that the collision of the policy worlds of security and public health has not produced a partnership of equals.

National and international public health authorities require two basic kinds of capabilities in order to address threats posed by pathogenic microbes – capabilities for surveillance and for interventions. Surveillance capabilities allow public health officials to know what disease threats are affecting what parts of what populations. Intervention capabilities let public health agencies act to prevent, protect, or contain disease threats. Stripped to its simplest form, public health governance seeks to develop and maintain sufficient surveillance and intervention capacities at national and international levels. The collision of the policy worlds of security and public health has underscored the governance importance of the surveillance and intervention functions.

Without question, the most important public health governance function is surveillance. As defined by the US Centers for Disease Control and Prevention, '[p]ublic health surveillance is the ongoing, systematic collection, analysis, interpretation, and dissemination of data regarding a health-related event for use in public health action to reduce morbidity and mortality and to improve health'.[25] Preventing, containing, or eliminating disease threats are impossible tasks if threats cannot be recognized and reported to public health authorities for action in a timely manner. The policy world of security can relate to the importance of surveillance for public health governance because it needs a similar capability – intelligence. Preventing and protecting against security threats are impossible missions if a government does not have intelligence about what security threats exist and the extent of their severity.

With threats identified, reported, and analyzed, public health officials can, if necessary, intervene against the threats. Public health governance involves three basic types of interventions (Figure 11.1). First, interventions can prevent disease threats from affecting populations. The objective of these interventions is to prevent pathogenic microbes from reaching the population. One example of a prevention intervention is the treatment of water supplies to ensure that pathogens do not reach people. Eradication campaigns are also prevention interventions because successful eradication

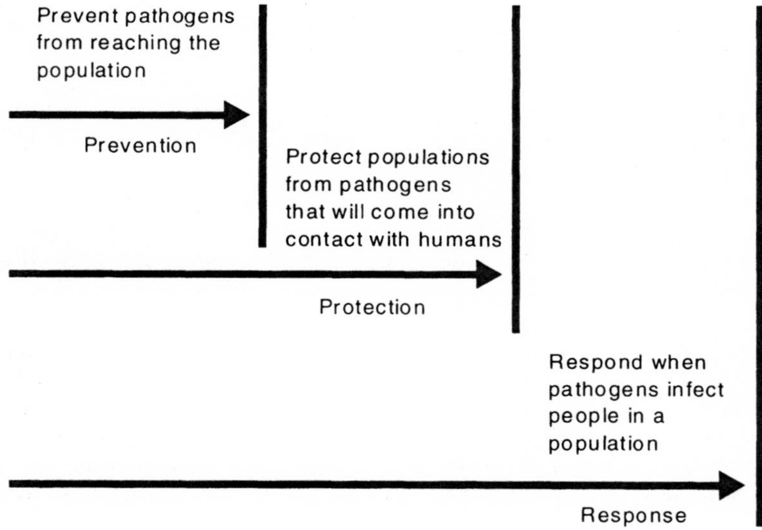

Figure 11.1 Types of public health interventions

of a disease will mean that no one comes into contact with it. Thus, the eradication of smallpox in the late 1970s has prevented millions of people from being exposed to smallpox.

Second, interventions can seek to protect people from pathogenic microbes with which they might come into contact. Protection interventions can include something as simple as hand washing or as sophisticated as a vaccine. Handwashing is a tried-and-tested intervention that reduces the likelihood that a pathogen in the environment will infect the person washing his or her hands. Vaccination against influenza also represents a protection intervention because the vaccine affords the recipient some protection against the influenza virus circulating seasonally in the community. Protection interventions in the context of HIV/AIDS include behavior change interventions, such as the promotion of abstinence and 'safer sex'. The premise of protection interventions is that people will come into contact with germs but that protection interventions 'harden the target' against incoming microbes.

Third, response interventions occur when a microbe infects a population, causing morbidity and mortality. Such interventions seek to contain infection by breaking the chain of transmission in the affected population and to mitigate the infection's impact on victims. Response interventions can involve complex technologies, such as anti-microbial drugs, or low-tech methods such as quarantine and isolation. By bringing infection under control, response interventions allow public health to undertake, if possible, prevention and protection interventions to prevent similar outbreaks in the future.

Although crude, this description of surveillance and intervention as the critical public health governance functions reveals two further points. First, effective surveillance and interventions require a significant commitment of trained personnel, economic resources, and sustained political understanding of the importance of public health to a society. Second, surveillance and intervention are interdependent governance functions. Weak surveillance undermines the effectiveness of any type of intervention. Likewise, inadequate intervention squanders information provided by a surveillance system. This interdependence makes public health governance fragile should political and economic support for public health wane.

Biosecurity, public health governance, and the synergy thesis

The emergence of biosecurity has, without question, focused increased attention on public health surveillance and intervention capabilities at national and international levels. Addressing the threats posed by bioweapons and naturally occurring infectious diseases requires robust surveillance and intervention capacities. Framing both threats as security challenges makes these public health governance functions more important politically than they had previously been. Security officials and public health experts reached the same conclusions about the condition of national and international public health – significant improvements would be required for biosecurity to be achieved.

One feature of convergence in thinking about biosecurity is the 'synergy' or 'dual purpose' thesis. This thesis posits that improvements made to public health would benefit efforts against both bioweapons and infectious diseases. When an outbreak occurs, experts argue, the first line of defense is the public health system, whether the outbreak was intentionally caused or naturally occurring.[26] Thus, strengthening public health through biosecurity policy would achieve the dual purpose of defending against bioweapons and protecting societies from naturally occurring infectious diseases. An implication of the synergy thesis is that biosecurity policy does not have to choose between security and public health or between defense against bioterrorism and protection from infectious diseases.

As subsequent sections of the chapter examine, the dual purpose or synergy thesis has merit but not as much as many people believed in the early stages of the development of biosecurity strategies. The synergy thesis proves more accurate in connection with surveillance than with interventions. Further, the synergy thesis does not capture tensions or frictions that exist between policies designed to deal with bioterrorism and those needed to address resurgent infectious diseases. The synergy thesis became biosecurity's equivalent of the 'harmony of interests' doctrine familiar to international relations experts.[27] This doctrine described how states, especially powerful countries, would maintain that a harmony existed between their interests and the interests of other countries or the 'international community'. The reality of international politics usually reveals not a harmony but a divergence of interests. Similarly, the reality of biosecurity suggests that the synergy thesis is insufficient to understand the challenges created by coupling security and public health together to respond to threats from bioweapons and infectious diseases.

Biosecurity and surveillance strategies

The importance of surveillance to both defense against bioweapons and infectious diseases has been recognized. Real-world crises, including the anthrax attacks against the United States and the global SARS outbreak, hammered home the critical nature of surveillance in both contexts. With both security and public health stressing the importance of surveillance, this endeavor has received more attention and resources than ever before. In the wake of the anthrax attacks, the United States strengthened public health surveillance within its territory and recognized that global surveillance played a role in the nation's strategy to counter the bioterrorist threat. Improving surveillance at the global level has been a key theme of most bodies and experts making recommendations about how to manage resurgent infectious diseases. WHO began the improvement process in the latter half of the 1990s with the development and implementation of its Global Outbreak Alert and Response Network (GOARN). For WHO, GOARN's capabilities bolster defenses against bioterrorism and naturally occurring infectious diseases, and are central to its strategy of achieving global health security.

The nature of surveillance as a governance activity lessens the potential for the fault line between bioweapons and naturally occurring infectious diseases to damage biosecurity policy. Improving the speed, comprehensiveness, and technological capabilities of surveillance is a 'dual purpose' objective of biosecurity policy. Synergy between security and public health interests on surveillance tends, thus, to be robust. The most frequent criticism made concerning surveillance is that, despite progress, national and global surveillance capabilities still fall far short of what biosecurity requires. Two reports on US bioterrorism policies published in December 2004 both noted the inadequacy of US surveillance capabilities.[28] Globally, WHO remains concerned about infectious disease surveillance and has succeeded in ensuring that the revised International Health Regulations include a duty for states' parties to develop and maintain core surveillance capacities to detect disease events and report public health emergencies of international concern to WHO.[29]

The recognized inadequacy of national and global surveillance systems raises questions for surveillance initiatives that do not build dual purpose capabilities for both security and public health. The United States has, for example, launched BioWatch, a $100 million effort to place environmental sensors in 30 cities to detect possible bioweapons attacks against the United States.[30] The idea behind BioWatch is to create capabilities to detect a possible bioweapons attack earlier than public health surveillance systems, thus giving response and recovery activities more time to be effective. BioWatch's sensors do not, however, promise to add much, if anything, to public health surveillance because they focus on detecting very low-probability releases of bioweapons agents. In other words, security concerns about bioweapons have created a surveillance program offering little benefit to public health's task of addressing infectious diseases nationally and internationally. Public health governance might more appropriately target the millions spent on BioWatch on the serious problems still plaguing infectious disease surveillance generally in the United States and internationally.

Biosecurity's other fault line – that between national and international focus – appears more starkly in connection with surveillance than the fault line between bioweapons and infectious diseases. Again, the robustness of the synergy thesis in the governance function of surveillance means, generally, that improvements in surveillance nationally will benefit surveillance capabilities internationally. The main reason for this situation is the division of humanity into approximately 190 sovereign states. International surveillance systems are, in large part, the sum of national surveillance capabilities. This reality is why WHO wanted states parties to the revised International Health Regulations to be under an obligation to improve their national surveillance capacities. The problem emerging from this dynamic is the growing 'surveillance gap' between developed and developing countries, a gap perhaps widened and deepened by the rise of biosecurity.

The surveillance gap simply describes the growing gulf between the ability of developed and developing countries to conduct infectious disease surveillance. This gap forms part of public health disparities between rich and poor countries, especially disparities concerning infectious disease morbidity and mortality. Most developing countries need improved surveillance for general public health purposes not security against bioweapons attacks, and these countries need financial and other forms of assistance to improve their surveillance abilities. Yet, developed countries are spending significant sums on improving their own surveillance systems not those of countries in the developing world. The BioWatch initiative helps illustrate how the current direction of biosecurity policy in the United States emphasizes narrowly conceived national interests over more broadly-framed global needs. BioWatch sensors in US cities sniffing the air for bioweapons agents will not help any country, including the United States, detect the emergence of new, virulent pathogens in the developing world with the potential to spread globally through international trade and travel. The widening surveillance gap reinforces the public health governance warning raised above about biosecurity policy disproportionately focusing on fear of bioweapons at the expense of broad-based improvements in surveillance of infectious disease events of whatever origin.

Biosecurity and intervention strategies

The second pillar of public health governance is the ability to intervene to prevent, protect against, or respond to a serious infectious disease threat. Intervention strategies have received significant attention in the development of biosecurity policies; and, as with surveillance, experts often stress the synergies achieved by developing intervention capabilities for either area of biosecurity concern. For example, emergency response plans developed for a bioterrorist attack using smallpox may prove beneficial in the event of pandemic influenza, and vice versa. Similarly, anti-virals stockpiled for a bioweapons attack may be used to address an unforeseen epidemic triggered by the emergence and spread of a new virus.

The fault lines of biosecurity policy are, however, more prominent concerning intervention strategies than surveillance. Part of this prominence reflects the diversity of intervention strategies that exist. As Figure 11.1 depicted, prevention, protection, and response interventions are distinct categories; and, within each category, multiple approaches exist. Such complexity means that some intervention strategies relevant to infectious diseases may have little relevance for defense against bioweapons, and vice versa. Despite its fondness for abstinence as an intervention for stopping the transmission of HIV/AIDS, the Bush administration is unlikely to integrate abstinence into its approach to bioweapons because this public health intervention makes no sense in defending against bioweapons attacks. Similarly, the complexity of intervention as a component of public health governance highlights tensions that may arise between national and international approaches. Is a country's national vaccine stockpile to be shared with other countries in the event of an international emergency?

The difficulty of creating robust 'dual purpose' approaches to biosecurity in the intervention area can be illustrated by comparing prevention interventions for bioweapons and infectious diseases. One strategy used to prevent pathogens from infecting populations has been disease eradication through vaccination, as happened with smallpox and as WHO is attempting with polio. Eradication does not, however, have much relevance for preventing bioweapons agents from threatening populations. In fact, disease eradication can create a bioweapon opportunity because of the vulnerability of unvaccinated populations. For example, smallpox's eradication has made it a feared bioweapon agent.[31] Similar worries accompany the potential eradication of polio, creating the need for keeping stocks of the polio virus in laboratories secure from theft and misuse.

The prevention objective for bioweapons has to be to prevent bioweapons agents from being obtained and developed by states or terrorist organizations potentially interested in biological mayhem. This objective requires regulating and controlling the storage and transfer of dangerous pathogens; the safety and security of laboratories in which such pathogens are kept; vetting the backgrounds of scientists and technicians who work with, or in facilities housing, bioweapons agents; and overseeing scientific research done on dangerous pathogens. Many of these tasks are familiar to national military bioweapons research but not to national and international civilian research, which was not subject to extensive regulation for security purposes until concerns about bioterrorism achieved critical mass in the latter half of the 1990s. Civilian research in the biological sciences has not experienced the security oversight under which civilian nuclear research historically proceeded. The new context of biosecurity creates unprecedented public health governance challenges that connect to the creation of scientific knowledge, its dissemination, and use in innovative health technologies.

Ensuring pathogen, laboratory, scientist, and research security with respect to dangerous pathogens delivers benefits for public health because it helps prevent intentional or accidental releases of germs with outbreak potency. This synergy comes, however, at a price in three respects from a public health perspective. First, increasing pathogen, laboratory, scientist, and research security means creating a regulatory and oversight system that requires significant resources to sustain effectively. From public health's perspective, these resources represent an 'opportunity cost' because the money spent on these aspects of biosecurity cannot be used to address other problems badly in need of funds.

Second, other biosecurity policies could increase these costs substantially, such as policies that expand scientific research into dangerous pathogens for the purpose of developing countermeasures (for example drugs and vaccines) against bioweapons agents. The United States is significantly expanding research into dangerous pathogens considered likely bioweapons agents, through initiatives such as Project BioShield.[32] This policy direction has caused controversy, with critics concerned that the expanded research program and additional facilities (1) are not justified by the low probability of bioweapons attacks; (2) will make it more likely that dangerous pathogens fall into the wrong hands or cause public health crises through failures in laboratory safety and security; and (3) will drain funding away from other areas of scientific and public health research.[33] Laboratory-contracted cases of tularemia – a recognized bioweapons agent – at Boston University in 2004 (but not reported to public health authorities in a timely manner) have highlighted criticisms of the manner in which the United States is expanding research into dangerous pathogens.

Third, intensifying research into dangerous pathogens for biosecurity purposes raises the ambiguous line between 'defensive' and 'offensive' research with respect to bioweapons. Concerns have been raised that some biosecurity research planned by the United States may violate, or raise the perception that the United States is violating, the BWC.[34] The United States defends novel research on dangerous pathogens as necessary to anticipate how states or terrorists may bio-engineer pathogens to overcome established defenses and countermeasures.[35] Such research constitutes a form of pre-emptive biodefense by the United States. This controversy demonstrates that the fault line between national and international interests in biosecurity appears with respect to increasing research on dangerous pathogens.

The category of protection interventions does not fit the security and public health components of biosecurity equally. Public health uses many protection interventions (for example, vaccines, education, behavior modification techniques) because it assumes that populations will come into contact with pathogenic microbes. The policy world of security cannot make the same assumption in connection with bioweapons agents, except concerning military forces that might be subject to bioweapons attacks. Thus, protection interventions made by public health, such as influenza vaccination, produce no synergies for biosecurity with respect to defending against bioweapons agents. Similarly, military vaccination of troops against smallpox and anthrax to protect them from intentional release by enemy forces, has no 'dual purpose' potential for protecting against naturally occurring infectious diseases. Protection interventions for security and public health proceed, thus, as parallel rather than converging governance tasks for biosecurity policy.

The attempt in the United States to vaccinate public health and health care personnel for smallpox reveals the extent to which prevention interventions do not work well with respect to bioweapons agents. The idea behind vaccinating individuals who would represent the 'troops in the trenches' in case of a smallpox attack on the US population was to protect such essential 'troops' in the time of crisis – essentially the same reason the military has vaccinated troops for anthrax and smallpox. The smallpox vaccination campaign, which in its first phase aimed to vaccinate 500,000 people, failed to achieve a vaccination rate anywhere near the original goals because an overwhelming proportion of people targeted for vaccination decided not to be immunized because of the health risks involved.[36] A similar fate probably awaits vaccines developed in the future as countermeasures to bioweapons agents, which means that such vaccines will function as a response intervention.

The deepest synergy levels for biosecurity policy in intervention appear in capabilities for responses to infectious disease outbreaks. Both security and public health experts acknowledge that addressing intentionally-caused or naturally-occurring outbreaks requires similar types of interventions to control the epidemic, break any chain of transmission, treat the infected, and move society toward recovery. Such interventions include mass distribution and administration of antibiotics or vaccines, surge capacity in health care facilities to provide clinical treatment to the ill, and the capability to engage in quarantine and isolation of infected persons or those suspected of being infected. Training exercises, such as the bioterrorism-focused TOPOFF ('Top Officials') exercises in the United States, and real-world outbreaks, such as anthrax and SARS, serve as evidence that even sophisticated public health and health care systems in developed countries are unprepared for the challenges severe infectious disease outbreaks would create.

The fault lines of biosecurity policy appear strongly in the area of response interventions, which means that the governance task of improving response capabilities faces difficult choices and a potentially contentious political context. Choices being made in the United States about response interventions illustrate the tensions that arise between bioweapons and naturally occurring infectious diseases in this area. The US government is committing billions of dollars to initiatives designed to provide public health and health care personnel with drugs and vaccines to respond to bioweapons attacks. These initiatives include a strategic stockpile of medicines for treating victims of bioterror attacks, a stockpile of smallpox vaccine with enough vaccine for every US citizen, and Project BioShield's effort to encourage the development and acquisition of new vaccine and drug countermeasures against bioweapons. With the exception of anti-viral treatments and medical supplies in the strategic stockpile, none of these initiatives provide 'dual purpose' capabilities for the United States in the event of pandemic influenza, considered by many public health specialists to be far more likely than another bioterrorist attack. Nor do these initiatives deliver benefits to the on-going global struggle to contain the spread of, and devastation caused by, HIV/AIDS. In the United States, the sums being directed at countermeasures against bioterrorism ($7.6 billion requested from Congress for fiscal year 2005) substantially exceed annual US contributions to the global fight against HIV/AIDS.

The national/international fault line also appears in the area of response interventions. While improved national capabilities to respond to severe infectious disease outbreaks contribute to international preparedness, policies that almost exclusively focus on national interests in biosecurity preparedness do not significantly help efforts required globally to increase response intervention capacities. Thus, biosecurity policies run the risk of increasing the 'response gap' that exists between developed and developing countries on preparedness to manage severe infectious disease outbreaks. For much of the world, naturally occurring infectious diseases, not bioweapons attacks, are the greatest source of concern with respect to response preparedness. In terms of response interventions, both the bioweapons/infectious disease and national/international fault lines generate serious governance tensions for biosecurity policy that must be effectively managed.

Emergence of a 'biosecurity industrial complex'?

The effective management of the governance fault lines appearing in biosecurity policy may be made more difficult if current trends in biosecurity policy are not reassessed. The momentum in US biosecurity policy that privileges bioweapons over naturally occurring infectious diseases and national over international interests may groove developments in this area in a way that deepens these fault lines and makes policy recalibration and reform more difficult. We could see a 'biosecurity industrial complex' emerge in which government bureaucracies, academic institutions, and private-sector enterprises develop entrenched interests in the continuation and acceleration of present policy directions.

The dramatic increase in US government attention on, and appropriations for, the threat of bioweapons has already produced a 'biosecurity policy industry' that encompasses federal and state governments, universities, and the private sector. Calls for even more spending on national defense against bioweapons, such as Senator Bill Frist's argument in January 2005 for 'something that even dwarfs the Manhattan Project' for US biodefense, indicate that biosecurity may continue to be a growth industry for governmental and non-governmental actors and interests.[37] The money spent on US biodefense will fund the creation of bureaucratic turf (for example, Department of Homeland Security), academic institutions and reputations (for example, universities seeking to build BSL-3 and BSL-4 laboratories to do biodefense work),[38] and potential profit streams for pharmaceutical and high-technology companies (for example, economic incentives created by Project BioShield and Project BioWatch), all of which will provide incentives for people invested in biosecurity policy to argue that continued, and even more, funding is required to protect the United States from bioterrorist attack. The 'biosecurity industrial complex' may develop as these interests seek to influence government decisions about the amount and kind of spending that is required to keep the United States safe from bioterrorism.

Concerns are appearing about the narrow focus of US biosecurity policy and the dynamics it may be creating. For example, a group of 758 microbiologists in the United States wrote an open letter to the National Institutes of Health (NIH) in March 2005 arguing that '[t]he diversion of research funds from projects of high public-health importance to projects of high biodefense but low public-health importance represents a misdirection of NIH priorities and a crisis for NIH-supported microbiological research'.[39] Similarly, experts criticized a highly publicized smallpox outbreak exercise, called Atlantic Storm, conducted in January 2005 because they believed that the exercise involved implausible assumptions predicated on worst-case scenario analysis and thus overstated the threat for political purposes.[40] Both episodes demonstrate that the fault lines analyzed above are real and becoming more entrenched as US biosecurity policy funds bureaucracies, institutions, and companies dependent on the present policy trajectory. Thus, we may be witnessing the emergence of a biosecurity industrial complex that could make needed policy reforms more difficult in the future.

The double burden of biosecurity on public health governance

Analyzing biosecurity's implications for surveillance and intervention reveals a double burden biosecurity imposes on public health governance. The first burden is conceptual. The collision of the policy worlds of security and public health has not produced a partnership of equals because the collision was not between policy equals. The emphasis on national defenses against bioweapons in US biosecurity policy reveals the continuing strength of traditional conceptions of security that focus on defending the nation from exogenous attacks against its power and material capabilities. The rise of biosecurity has encouraged some experts to posit that traditional notions of security are fading in the face of more expansive perspectives that resonate better with public health's traditional concerns with human health in a globalized world. The conceptual transformation represented by biosecurity may, however, reveal a new type of subordination of public health to security. The previous hierarchy, captured by the distinction between the 'high politics' of security and the 'low politics' of public health, was a distant one because security and public health did not deeply intersect in their respective historical pursuits. The current hierarchy is direct and immediate because security and public health have become intertwined areas of national and international policy; the hierarchy is now one within the realm of 'high politics'. This change elevates public health conceptually but perhaps, ironically, reinforces its subordination to security as a governance pursuit.

The second burden biosecurity creates for public health governance is practical. The political importance and range of surveillance and interventions activities now falling under the rubric of public health governance have increased significantly as biosecurity policy has developed. Public health governance now has to navigate the shoals of both bioweapons and naturally occurring infectious diseases at national and international levels informed by security and public health concepts and communities. This chapter has only mentioned a handful of the many and diverse challenges biosecurity policy confronts. In addition, public health governance still bears the burden of addressing public health threats that do not constitute biosecurity threats, such as growing national and international problems with non-communicable diseases. This taxing, dangerous, and stressful environment in which public health now operates is not an ephemeral sea change but an upheaval of epic proportions that will change the theory and practice of public health for generations.

The double burden biosecurity imposes on public health governance suggests that public health should see biosecurity neither as friend nor foe but as a dominating fixture in its current and future policy landscape. The upheavals that have made biosecurity a fixture for public health governance are so recent that attempting to reach definitive conclusions about how exactly biosecurity will shape public health governance, and whether such governance will likewise shape biosecurity, is a fool's errand. Biosecurity is largely the product of reactions to crises anticipated and unforeseen; and future crises may well determine the relationship between biosecurity policy and public health governance more than sincere, reflective efforts to prepare and plan for future challenges.

Conclusion: Biosecurity, public health, and good governance

This chapter has focused on the transformation of public health governance created by the emergence of biosecurity as a policy endeavor. This transformation contains sobering implications for public health that require acknowledgment if public health governance is to find its way in this new policy environment. However, the sobering elements for public health found in the rise of biosecurity should be placed within a larger framework of analysis that considers the role of public health in 'good governance' in the twenty-first century. Conceptions of good governance have not typically identified public health as a key ingredient of such governance, preferring instead to focus on ideas such as democracy and the rule of law. The role of public health in biosecurity helps reveal, however, how important public health is becoming for good governance nationally and globally.

The elevated political importance of public health seen in the development of biosecurity policies also features in governance contexts other than security, namely economic prosperity, development, and protecting and promoting human dignity. The post-Cold War world has witnessed public health's growing significance for international economics. The various controversies over the World Trade Organization's impact on national and global public health signal a shift in how public health relates to international economic governance. Further, public health has emerged as a central objective in development thinking, as illustrated by the recommendations made the WHO's Commission on Macroeconomics and Health and the health objectives and targets contained in the UN's Millennium Development Goals. Finally, public health has also emerged as an important issue in the human rights area with respect to both civil and political rights and economic, social, and cultural rights. The heightened importance of public health in human rights resonates with public health's traditional connections to the promotion and protection of human dignity.

The importance of public health across security, economic, development, and human rights agendas suggests that it is becoming an independent marker of good governance in the early twenty-first century. This broader context helps keep public health's collision with the world of security in perspective and provides incentives for those working in public health to engage fully with this emerging new world order. The fault lines of biosecurity policy demonstrate that making public health an effective component of good governance nationally and globally will not be easy, and success is not inevitable. But, never before has public health been as important to security and other fundamental governance objectives. Biosecurity has transformed public health, but it has also produced the opportunity for public health to participate in the broader governance transformations affecting virtually every area of human endeavor as the twenty-first century unfolds.

Notes

1 Office of Homeland Security, *National Strategy for Homeland Security* (Washington, DC: The White House, 2002).
2 For discussion, see D.P. Fidler, 'Public Health and National Security in the Global Age: Infectious Diseases, Bioterrorism, and *Realpolitik*', *George Washington International Law Review*, 35 (2003): 787; and D.P. Fidler, 'Fighting the Axis of Illness: HIV/AIDS, Human Rights, and US Foreign Policy', *Harvard Human Rights Journal*, 17 (2004): 99.
3 M.K. Kindhauser (ed.), *Global Defence Against the Infectious Disease Threat* (Geneva: WHO, 2002).
4 J.F. Hutchinson, *Champions of Charity: War and the Rise of the Red Cross* (Boulder, CO: Westview Press, 1996), p. 126.
5 Medical Museum of the Armed Forces Institute of Pathology, *Yellow Fever* (Washington, DC: Armed Forces Institute of Pathology, 1964), p. 1. See also Stern, this volume, chapter 7.
6 P.H. Manson-Bahr, *History of the School of Tropical Medicine in London* (London: Lewis, 1956), p. 31 (observing that, in the late 1890s, British political officials and physicians realized that '[i]nstruction in tropical medicine was urgently necessary in this country as it was the centre of a great and growing tropical Empire').
7 Military efforts to address venereal disease had, for example, consequences for the spread of such diseases in civilian populations. See L.A. Hall, '"War Always Brings It On": War, STDs, the Military, and the Civilian Population in Britain, 1850–1950' in R. Cooter, M. Harrison and S. Sturdy (eds), *Medicine and Modern Warfare* (Amsterdam: Rodopi, 1999), p. 205.
8 P.J. Weindling, *Epidemics and Genocide in Eastern Europe 1890–1945* (Oxford: Oxford University Press, 2000), pp. 51–2, 58.
9 Geneva Protocol for the Prohibition of the Use in War of Asphyxiating, Poisonous or Other Gases, and of Bacteriological Methods of Warfare, *League of Nations Treaty Series*, 44 (17 June 1925): 65.
10 Convention on the Prohibition of the Development, Production, and Stockpiling of Bacteriological (Biological) and Toxin Weapons and on Their Destruction, *International Legal Materials*, 11 (10 April 1972): 309.
11 J. Miller, S. Engelberg and W. Broad, *Germs: Biological Weapons and America's Secret War* (New York: Simon & Schuster, 2001), pp. 62–3.
12 E. Fee and T.M. Brown, 'Preemptive Biopreparedness: Can We Learn Anything from History?', *American Journal of Public Health*, 91 (2001): 721.
13 On the history of these efforts, see N.M. Goodman, *International Health Organizations and Their Work*, 2nd edn (London: Churchill Livingstone, 1971) and N. Howard-Jones, *The Scientific Background of the International Sanitary Conferences 1851–1938* (Geneva: WHO, 1975).
14 Constitution of the World Health Organization, 22 July 1946, in World Health Organization, *Basic Documents* 40th edn (Geneva: WHO, 1994), pp. 1–18.
15 *Ibid.*, p. 1.
16 Declaration of Alma Ata, in *Report of the International Conference on Primary Health Care* (Geneva: WHO, 1978).
17 See, for example, Institute of Medicine, *Microbial Threats to Health: Emergence, Detection, and Response*, M.S. Smolinski, M.A. Hamburg and J. Lederberg (eds) (Washington, DC: National Academy Press, 2003).
18 For a seminal analysis of human security, see United Nations Development Programme, *Human Development Report 1994* (New York: UNDP, 1994), p. 22.
19 United Nations, *Report of the Secretary-General's High-Level Panel on Threats, Challenges, and Change – A More Secure World: Our Shared Responsibility* (New York: UN, 2004).

20 *Ibid.*, p. 2.
21 *Ibid.*, p. viii.
22 *Report of the Secretary-General – In Larger Freedom: Towards Development, Security and Human Rights for All*, UN Doc. A/59/2005, 21 March 2005, pp. 20–1, 25–9. See also K. Annan, '"In Larger Freedom": Decision Time at the UN', *Foreign Affairs* (May/June 2005), p. 63.
23 D.P. Fidler, 'Health as Foreign Policy: Between Power and Principle', *Seton Hall Journal of Diplomacy and International Relations* 6 (2005): 179–94.
24 Fidler, 'Public Health and National Security', 833–56.
25 US Centers for Disease Control and Prevention, 'Updated Guidelines for Evaluating Public Health Surveillance Systems: Recommendations from the Guidelines Working Group', 50 *Morbidity and Mortality Weekly Report* (27 July 2001), p. 2.
26 L. Garrett, *Betrayal of Trust: The Collapse of Global Public Health* (New York: Hyperion, 2000), pp. 584–5.
27 See, for example, E.H. Carr, *The Twenty-Years' Crisis, 1919–1939: An Introduction to the Study of International Relations* (London: Macmillan, 1939), pp. 41–62.
28 Trust for America's Health, *Ready or Not? Protecting the Public's Health in the Age of Bioterrorism* (Washington, DC: Trust for America's Health, 2004). Chemical and Biological Arms Control Institute, *Fighting Bioterrorism: Tracking and Assessing US Government Programs* (Washington, DC: CBACI, 2004).
29 World Health Organization, *Revision of the International Health Regulations*, WHA 58.3, 23 May 2005, Articles 5.1 (surveillance capabilities) and 13.1 (response capabilities).
30 The White House, *Biodefense for the 21st Century* (Washington, DC: The White House, 2004), p. 2.
31 See generally D. Koplow, *Smallpox: The Fight to Eradicate a Global Scourge* (Berkeley: University of California Press, 2003).
32 The White House President Details Project Bioshield, 2 February 2003. http://www.whitehouse.gov/news/releases/2003/02/20030203.html (accessed 10 August 2006).
33 See, for example, H.W. Cohen, R.M. Gould and V.W. Sidel, 'The Pitfalls of Bioterrorism Preparedness: The Anthrax and Smallpox Experiences', *American Journal of Public Health* 94 (2004): 1667–71.
34 L. Goodman, 'Biodefense Cost and Consequence', *Journal of Clinical Investigation* 114 (2004): 2–3.
35 *Ibid.*
36 D. MacKenzie, 'US Smallpox Vaccination Plan Grinds to a Halt', *New Scientist*, (22 August 2003). http://www.newscientist.com/article.ns?id=dn4074 (accessed 10 August 2006).
37 B. Hirschler, 'Call for New Manhattan Project to Fight Bioterror', Reuters, 27 January 2005.
38 One non-governmental organization has identified many universities and university-related institutions interested in building BSL-4 and BSL-3 laboratories in order to house research activities that support US biodefense activities. See Sunshine Project, 'Map of the US Biodefense Program: High Containment Labs and Other Facilities', at www.sunshine-project.org (accessed 10 August 2006).
39 'An Open Letter to Elias Zerhouni', *Science Online*, 4 March 2005. http://www.sciencemag.org/feature/misc/microbio/307_5714_1409c.pdf (accessed 10 August 2006).
40 D. Ruppe, 'Experts Question Merit of Recent Smallpox Exercise', Global Security Newswire, 9 March 2005. http://www.nti.org/d_newswire/issues/2005_3_9.html#77244457 (accessed 10 August 2006).

12

Postcard from Plaguetown: SARS and the Exoticization of Toronto

Carolyn Strange

'"Bad news travels like the plague. Good news doesn't travel well"'. While this epigram might have appeared in an advertising or marketing textbook, they were the words a Canadian politician chose to explain why the federal government sponsored a Toronto rock concert in the summer of 2003. As the Senator stated, 3.5 million dollars was a small price to pay for an event held to restore confidence in a city struck by SARS. The virus – first diagnosed in Toronto in March 2003 – had already claimed 42 lives; international media coverage of the outbreak had strangled the economy. While public health officials imposed quarantine and isolation measures to combat the spread of SARS bureaucrats and business leaders were equally active, treating the virus as an economic crisis caused by negative publicity. 'SARSstock', as locals dubbed the concert, was one of many events prescribed to repair and revitalize the city's image post-SARS. By drawing close to half a million fans with big name musicians, including The Rolling Stones, it provided Canadian and US newspapers and television outlets with a splashy 'good news' item. Local media commented that the concert gave Torontonians a much-needed tonic. As the Toronto *Star* declared, it proved to the world that 'life and business here rock on'.[1]

When the World Health Organization (WHO) issued its advisory against 'unnecessary travel' to Toronto in April 2003, not only suspected SARS carriers but the city itself felt borders close around it, cutting off vital flows of traffic to the city.[2] It is thus possible to frame the economic consequences of SARS within the longer history of quarantine, *cordons sanitaires* and their commercial dimensions (as Hooker also argues in Chapter 10). Difficult to police and frequently a flashpoint for violent protests, public health measures that restrict the movement of people and goods have typically spawned economic crises. Indeed, distress over the financial impact of disease containment, rather than concerns over prejudicial treatment of suspected carriers, have historically pushed medical authorities to relax quarantine.[3] While strict maritime quarantine provisions gave way over the nineteenth century to individual medical inspection, a governing strategy that focused on monitoring individual movement, quarantine continued to clash with commerce in the twentieth century, as Nayan Shah and others have shown.[4]

In Toronto, however, the SARS-induced economic crisis sprang primarily from a mediatized culture of fear and externally-imposed travel advisories, rather than from locally-imposed quarantine regimes. Alarming news stories about the virus's transmission and deadliness chilled tourism, Toronto's second-largest industry (after banking and finance), and a sector largely dependent on image.[5] In the midst of the crisis (at which point 16 people had already died) Toronto Mayor Mel Lastman announced that the city's publicity problem overshadowed its health emergency: 'it's not the disease that's doing the damage – it['']s public perception about SARS that's hurting Toronto's tourism industry'.[6] This statement announced the city's commitment to find ways to reinvent its identity – a campaign that would prove as daunting as the effort to combat the virus.

How did SARS recast Toronto in the eyes of the world, and how was its image refashioned into one that signaled the renewed vitality of life and business? This post-modern public health crisis unfolded in ways that took Torontonians by surprise. First, while local health bodies ordered individuals into isolation it was not city, provincial or federal authorities that imposed travel restrictions but the supra-national WHO. Twenty-six countries followed suit with their own travel warnings, while many businesses and professional associations advised their employees and members to cancel Toronto trips (or in many cases, any travel to Canada). This, then, was a moment when a global body undermined the nation-state's capacity to regulate its own borders in the interest of commerce. Second, international media coverage of SARS transformed Toronto into a latter-day plaguetown, an economically vibrant modern city cut to its knees by a mystery bug.[7] Not only was Toronto sick, but newsmakers identified SARS as an *exotic* virus, transported by an Asian person from Hong Kong. The new definition of Toronto as 'the only place outside of Asia' to be placed on the WHO's travel advisory list was more than a statement of fact; repeated endlessly the phrase became a trope that merged Toronto with Asia, on the one hand, and underlined its anomalous inclusion, on the other. These multiple processes of identification textually exoticized Toronto, by tying it authoritatively (confirmed once epidemiologists traced SARS to southern China) to the 'East'. Knocked off its secure Western perch, the North American city fell into the imagined geography of dangerous disease, unexpectedly transmuted into an object of curiosity and fearful scrutiny.

Recovering from SARS, particularly through state-sponsored tourism promotion, such as the SARSstock concert, involved far more than proclaiming that the virus had been conquered; rather, as I argue, these efforts exposed, confronted and courted ambivalent sentiments about race in an officially multicultural city. Promotional schemes worked through a variety of media to recast the city as safe and as safely Western. Official strategies designed to address Toronto's 'publicity problem', including promotional songs, entertainment packages, postcards, television commercials, websites, and magazine advertisements, worked initially to de-exoticize the city, by inducing Torontonians to become 'tourists in their own town'. The second phase looked outward and drew on Toronto's pre-existing identity as an ethnically and culturally diverse city, with colorful and consumable ethnic enclaves.[8] Once SARS seemed defeated, tourism marketers gave the city's exoticism a romantic and commercially appealing gloss, free of dangerous associations.

Toronto, exotic plaguetown

SARS constituted a medical crisis but it also sparked a historically-coded and racially-loaded struggle over representation. Multimedia representations of the SARS outbreak in Toronto incorporated the city within the code of medieval plague. In March 2003 SARS the virus began to attack human bodies; at the same time SARS, the new brand name for 'mysterious deadly infectious disease', marked Toronto as plaguetown. 'SARS' signified the new millennium version of plague, in which an exotic disease 'invades' a modern city, thrusting its citizens into what one journalist called a 'new medievalism' of quarantine and fear.[9] Toronto's perplexing outbreak connected historical narratives to new doomsday imaginings of wealthy, modern cities' vulnerability to military and microbial threats.[10] Likened to the 9–11 attacks and the anthrax scare of 2001, the event inspired North American and European journalists to ponder: 'if it could happen in Toronto, was anywhere [in the affluent West, by implication] safe?'[11] While representations of SARS pitched Toronto into the past and linked it to the new threat of anti-Western terrorism, they also wrenched the city from its geographical moorings and seemingly secure Western identity. News reports of the disease's origins as a virus that had jumped from 'exotic' animals to humans in Guangzhou invoked Western cultural scripts of the mysterious and dangerous Far East.[12]

Analyzing the touristic and marketing aspects of SARS' impact on Toronto does not diminish the seriousness of the disease, which claimed a total of 44 lives. Aside from those who died, hundreds of people became seriously ill, and many have failed to recover fully.[13] Torontonians in isolation faced financial and personal hardship, while thousands of hotel and restaurant workers were laid off, many permanently. Thus images of Toronto – the sick city, its normally bustling streets eerily deserted – were hardly fictional. But the material effects of news discourse were powerful and pervasive; mass media narratives of Toronto as a city that had caught an exotic disease racialized SARS and authorized what Asian people defined as racial profiling.[14]

As Peter Mason argues, studying the exotic is to take as a subject of inquiry 'a representational effect, produced and reproduced', rather than a quality fixed in any geographic, ethnic or ontological sense.[15] While the exotic has historically referred to the non-Western 'other', conceptions of difference also surface in dominant constructions of familiar people or places, which appear strange or illogical. Thus when international public health experts, along with Western media outlets represented Toronto as a North American place that had caught an 'Asian' disease, same flipped into other, and the famously bland city became urbana incognita.

Several weeks prior to the diagnosis of SARS in Toronto, international health experts had traced the virus's origin to China. Already associated with mysteriousness and 'inscrutibility' in Western representations of 'the Orient', China's association with this new health threat reinforced historic Eurocentric readings of the Chinese as diseased and dangerous.[16] While travel promoters in non-Western parts of the world (and Asia and Africa in particular) still manipulate Western scripts of exoticism for profit, negative judgments bloom at the first sign of bad news, as they did when Western news outlets first reported that atypical pneumonia had emerged in southern China. Following the WHO's accusations that the Chinese government had been uncooperative and secretive, Western television and print media repeatedly broadcast images of Chinese sanitation workers scouring wild animal markets, wiping countertops with disinfectant, and posting 'no spitting' signs on lamp posts. And when the WHO issued its first-ever travel advisory on 2 April, recommending against non-essential trips to Hong Kong and Guandong Province, Western governments and news outlets readily accepted the judgment that East Asia ought to be avoided.

The WHO's second travel advisory was another matter. Its warning against non-essential trips to Beijing, Shanxi Province and *Toronto* performed a double function: on the one hand, Toronto was now officially one of the world's 'hot zones' of disease; on the other, the announcement erased the city's imagined and projected identity as a safe, North American city. WHO travel alerts in April and May, which referred exclusively to Asian destinations (Taiwan, Singapore, the Philippines, and further Chinese provinces) reconfirmed the city's newly exotic identity. Consequently Toronto, 'the only location outside of Asia' to have warranted a travel advisory, became strange within the West. Historic plague and leprosy references (as in, 'Toronto is limping toward "leper status"') were also racially coded and loaded ('Basically Toronto is taboo')[17] in print and in television news lead-ins. On the day when the WHO issued its advisory, for example, the Toronto *Globe and Mail's* front page read: 'SARS creates a city of pariahs. Shunned by Canada and the US alike, residents increasingly feel like lepers'. The story went on to add that 'US newspapers have nicknamed [Toronto] Pariah City'. These signifying practices, which deployed a non-

English term (pariah) for stigmatized, turned Toronto into a loathsome *and* dangerously exotic locale – a city, like lepers, that deserved to be shunned.

The scope and deadliness of SARS in Toronto, and doubts about the capacity of government and medical agencies to cope reinforced the city's image as a backward place, not quite up to modern standards of hygiene and scientific rationality. When SARS first appeared in Toronto in the spring of 2003, the virus' spread seemed incomprehensible, considering the city's wealth of medical facilities and expertise.[18] But when Toronto's hospitals turned out to be disease vectors rather than places of cure, and when its public health officials squabbled publicly, disseminating conflicting information about transmission and symptoms, Toronto's North American know-how lost all credibility. Even more embarrassing to Canadian health ministries was the fact that comparatively poor and under-resourced Asian countries, most notably Vietnam, managed to contain SARS more quickly and effectively.[19] The rapid spread of SARS in Toronto, and its reappearance several weeks after health officials officially declared the outbreak over left the city in company with the world's worst-hit locations – Guangzhou, Beijing and Hong Kong. Exoticized representations of Toronto thus drew upon and reinforced the lethal failures of its public health system.

If this is how Toronto appeared to the world, Canadian media coverage of SARS, in contrast, defined the virus as an exotic invasion. Without expressly accusing Chinese immigrants of transporting SARS news stories played on historic anti-Asian and anti-immigrant attitudes. In late-nineteenth-century Canada, as in many other countries where Chinese people had immigrated or sojourned as workers, immigration restrictions were implemented in the form of racially-specific head taxes. In 1923 the federal government passed the *Chinese Immigration Act,* more commonly known as the Chinese Exclusion Act, which all but prevented immigration of Chinese nationals, between 1923 and 1947. After this Act's repeal, and the institution of an official policy of multiculturalism in the 1970s, immigration from China and other parts of Asia increased substantially, comprising a key component of the Canadian 'mosaic'.[20] But shifting immigration patterns and multicultural flag-waving failed to erode deeply ingrained cultures of racism and xenophobia, which have reasserted themselves in moments of national anxiety. For example, in 1999 news accounts of illegal arrivals from China speculated before they had even arrived that they might be criminals (why had they not applied to immigrate legally?) and disease-carriers (they probably had AIDS); in both respects these Chinese were national security threats.[21] As the Chinese Canadian National Council protested, the SARS crisis in 2003 did not invent associations between the Chinese, disease and danger in the minds of Euro-Canadians; rather it ignited an existing store of anti-immigrant and anti-Chinese policy and sentiment in Canada generally and Toronto specifically.[22]

Content analysis of mainstream print media coverage of SARS confirms that Canadian readers were cued to read the virus as an Asian plague. Images, headlines and text tied mystery, danger and foreignness. In the Toronto *Globe and Mail*, for example, 44 per cent of its stories on SARS, between April and June 2003 included images of Asian people, particularly Asians wearing masks.[23] The racialization of SARS was anchored through the public identification of Canada's first SARS victim, a Toronto resident, who had caught the virus while staying at Hong Kong's Hotel Metropole. Local papers named the woman, revealed where she lived, and provided her immigrant status (she was a landed immigrant who had only recently moved from Hong Kong). Thus, not only did SARS not 'belong' to Canada but neither did Toronto's case zero.[24] Suspicions that all Chinese people, irrespective of immigrant status or travel habits, were disease carriers led Torontonians to boycott businesses in the city's downtown and satellite Chinatowns. When Asian people called the Toronto Public Health agency's SARS hotline they frequently complained that they had been victims of 'racial profiling'. In her report on the city's efforts to combat SARS, the Chief Medical Officer of Health confirmed their allegations:

> certain groups were particularly burdened by the pressures of intense public attention. The Chinese and East Asian communities were associated with the geographical origin of the outbreak and other 'hotspot' locations where SARS was present.[25]

Accounts of Toronto's vulnerability to a non-native virus thus interpolated biological narratives of invasion and cultural narratives of racial contagion. These constructions conformed to Western discourses of the 'Far East' while displaying specifically Canadian strains of Sinophobia. Unlike the Chinese immigrants who had historically been excluded from Canada, or the boat people who had been deported, Canadians recognized that this foreign disease could not be shut out. As one journalist darkly observed, 'an exotic animal market in Guandong Province is only hours away'.[26]

A deep sense of misidentification, and not simply concern about Toronto's economy, provoked Canadian health officials and politicians to express 'bewilderment and anger' over Toronto's inclusion on the WHO's travel advisory list. One Canadian policy analyst expressed his disgust, by complaining that the WHO's announcement 'ha[d] soiled this city's signature reputation for hygiene'.[27] The city's Chief Medical Officer of Health flatly rejected Toronto's linkage with Asia and its SARS crisis: 'To categorize us as close to Beijing or other parts of China I think is a gross misrepresentation of the facts'. Toronto microbiologist Donald Low, a leading authority on the nature of SARS and its containment, was far less diplomatic: 'It's a bunch of bullshit!' In a cooler moment Dr. Low alleged that the WHO's announcement was based on 'politics', not science: 'I think they're looking for another scapegoat. And I think we're the scapegoat'.[28] By distinguishing politics from microbiological 'evidence' he hinted that the WHO had chosen Toronto in order to add a token Western location to its list of Asian SARS 'hotspots'. The *Times* of London agreed: 'The situation [in Toronto] is not at all the same as in Beijing... it was included for reasons of political correctness, to balance the advice on China'.[29] In an unguarded moment, the Province's commissioner of public health protested, 'We're not some rinky-dink Third-World country'. After all, Toronto's 'public health infrastructure' must necessarily be superior to any in Asia.[30] The most explicit charge of anti-Western global health politics appeared in Canada's right-wing *National Post*, in which an editorialist demanded: 'Why did the WHO strike at Toronto? Was it a case of national ethnic profiling – wanting to lift the burden off China and Asia and drop it into the lap of a whitish North America?'[31]

These shrill responses to Toronto's undesirable exoticization invoked the myth that Canada is a white, technologically advanced and safe nation. But swearing, accusing the WHO of politicking scientific incompetence, as well as accusing the WHO of anti-'whitish' racism failed spectacularly to reassure the world that Toronto had SARS under control. The city's public relations nadir occurred when Mayor Mel Lastman appeared on CNN's *Larry King Live* program immediately after the advisory was announced. The leader of North America's fifth largest city, Lastman squandered an opportunity to articulate reasoned objections to the WHO's decision and to advise millions of US and Canadian viewers that Toronto remained a familiar, safe city for visitors and residents. Instead he responded with an apoplectic display of arrogance and ignorance about the WHO:

> They don't know what they're talking about. I don't know who this group is, I [sic] never heard of them before. I'd never seen them before. Who did they talk to? They've never even been to Toronto. They're located somewhere in Geneva.[32]

Relying on his trademark folksy style, the mayor provided millions of viewers with reason to believe that the WHO's assessment of Toronto was sound.

Managing the exotic

From a more considered position than the one he took in his disastrous CNN interview, Mayor Mel Lastman struck a SARS Advisory Task Force and gave it the mandate to re-brand Toronto. Business leaders and civic boosters fought back by peddling promotional packages for tourists; just as significant were their strategies to 're-brand' the city. Refashioning Toronto's image became an economic and political priority, involving all three levels of government as well as task forces, businesses and community groups. Unlike colonial Capetown or Bengal, where imperial governance found expression in overtly racial strategies of rule, or early twentieth-century San Francisco, where health officials could impose race-specific restrictions on movement in the name of public health, the association of SARS with Toronto's Asian communities in 2003 had no official authorization. In fact, officially-orchestrated image recovery would depend on the selective deployment of Toronto's pre-SARS identity as an ethnically diverse city, with one of North America's largest and most visible Asian communities. How, then, to maintain the touristic appeal of ethnic exoticism while undoing powerful, historically-rooted associations between Asians and disease? This was the challenge that politicians, bureaucrats, and marketing consultants confronted. The unwanted branding of Toronto as plaguetown, and its subsequent re-branding post-SARS is thus an opportunity to analyze the discursive relationship between illness and the exotic beyond historians' temporal and geographic focus on colonial contexts and colonized peoples.

Western representations of the exotic have never been wholly negative, of course.[33] Imperial adventure writing on Asia and Africa depicts intriguing, enlivening destinations and contemporary commercial and governmental travel promotion continues to bank on these historic associations. Places historically exoticized in colonial literature reappear two or three centuries later in glossy brochures as appealingly out of the ordinary for Western tourists. Consider, for example, *Exotic destinations for wheelchair travelers: hotel guide to the Orient: Hong Kong, Macau, Singapore, Taiwan, Thailand*.[34] Here we see a typical sign linkage – exotic-Orient – with specific countries beaded along a pan-Oriental string. The WHO's travel advisory list was almost identical, save for the geographically and culturally jarring, faintly absurd addition of 'Toronto'.[35]

In Asian and other non-Western countries, private businesses and tourism agencies manipulate colonialist representations of their exoticism in order to gain a market edge over Western destinations. 'Marketing the margins' has become a standard element of cultural self-fashioning in tourist promotion, as well as in post-colonial literature and art.[36] An example of commercially motivated practices of self-exoticization appears on the Tanzanian government's official website, which invites the visitor to 'Rediscover the wild, romantic Africa of your dreams'. Needless to say, the hailed 'you' is not the Tanzanian citizen but the wealthy adventurer in search of the exotic, even if accessed only through a web portal.[37] Tourism Toronto's 'World within a city' campaign, launched after the SARS crisis subsided, would follow the same strategy, carefully avoiding any mention of an 'Asian' disease while vigorously promoting the city's 'authentic' Chinatowns.

The image of a mysterious plague raging through a city run by incompetent politicians and defensive medical authorities drew world-wide media coverage but it was hardly the publicity that local business leaders would have wished for. Well before the initial outbreak of SARS, programs had been tabled to freshen Toronto's image and revive its sagging tourist industry. In 2000, for example, the municipal government adopted a resolution to 'position Toronto by telling ourselves and the world what a really great City Toronto is'. Two years later the Commissioner for Economic Development, Culture and Tourism had presented an elaborate plan for the development and marketing of a comprehensive '"brand" strategy' for the city, one designed to replace its 'unfocused and unexciting' image – clean and safe, and not much else.[38] A coalition of civic leaders, named the Toronto City Summit Alliance, had reached a similar conclusion, asserting that 'stronger promotion of the Toronto region as a visitor destination' required well-funded marketing.[39] Ironically it took SARS and the subsequent public relations debacle to push all three levels of government to pump money into promotions designed to 'raise Toronto's international profile' on the city's own terms: to combat fear, to inspire pride, and to entice tourists and investors. While medical workers tackled the virus in the spring of 2003, advertising and marketing firms turned to the parallel task of remaking Toronto's image.

The redomesticated city

Tourism advisors faced a daunting and unfamiliar challenge: how could a city, previously renowned for its safety and cleanliness, regain firm financial footing if its own citizens, let alone foreign travelers and investors, feared walking the streets, riding the subway, attending work or school? While risk is a prominent concern in the international tourism industry,[40] it has never been a key issue for Canadian tourism marketers and officials, other than those who deal with the dangers of avalanches or bears. In tourist Canada, risk is remote, while cities are generally considered secure for visitors. Toronto's tourism industry has traditionally rested on its reputation for visitor security, certainly in contrast to neighboring US cities, where reports of citizens and tourists victimized in drive-by shootings, muggings and carjackings are frequent. In the Toronto tourism commissioner's 2002 report on Toronto's most marketable features, for example, 'safest large metropolitan area in North America', and 'comprehensive health care' were noted as key attributes of 'Toronto the Good'.[41] Before SARS, Americans were more likely to associate Toronto and Canada with bland civility, rather than riskiness. For instance, a 2002 database search of US news stories on 'danger' produced 75 hits for Toronto, compared to 690 for Los Angeles and 480 for Chicago. One US journalist who covered the SARS outbreak commented sarcastically on the absurdity of Toronto's new reputation for riskiness: 'The SARS epidemic requires American newspaper readers to adjust to the novelty of seeing the words 'danger' and Toronto in the same sentence... Canada is a famous underperformer for the creation of interesting news'.[42] As it happened, US travelers adjusted rather quickly to the news that Toronto had become risky. By the end of April 2003, 14 per cent of American air travelers declared that they had canceled trips to the city out of fear, and big-name American movie stars, performers, and sports figures withdrew from scheduled Toronto appearances.

SARS-struck Toronto suddenly appeared on the international mediascape in the way that exotic locales typically appear in Western newscasts: as a dangerous, mismanaged place. While some local politicians spoke gloomily about the economic costs others considered how the international publicity machine might be harnessed to revive the tourist industry. As prominent City Councillor Olivia Chow reflected, 'we didn't want this kind of publicity but now that we have the world's attention we might as well use it to send a different message about our city'.[43] Medical officials' daily briefings about Canadian public health measures and favorable assessments from the US Center for Disease Control concerning the outbreak's management were inadequate; they provided scientific information rather than reassuring images. Messages capable of swaying locals and potential visitors required different messengers, messages, and media, City Council decided.

In response the SARS Advisory Task Force's initiatives, and other recovery efforts conducted during and shortly after the initial SARS outbreak focused initially on redomesticating the city. Clustered around the concept of 'belonging', events and promotions encouraged locals to re-embrace their own city. A campaign of strategic exoticization emerged only after the city seemed to be securely SARS-free. Reminiscent of Toronto's pre-existing slogan, 'Diversity our Strength', post-SARS promotional campaigns emerged under the trademarked logo, 'Toronto: the World Within a City' – a place rich in consumable culture while also 'safe and secure'.

Belonging in Toronto

Official SARS recovery efforts initially turned against Toronto's worldly qualities in an effort to de-exoticize Toronto's image. As a non-native disease introduced by a non-native, SARS did not belong *in* Toronto and it did not belong *to* Toronto; yet there were millions of Torontonians who *did* belong, but who feared mixing in public, especially in the city's Chinese business districts. Recognizing that attracting foreign visitors was unlikely while travel advisories were in effect, the first tourism promotion campaigns targeted city residents with money to spend and a desire to help prop up the economy. In launching his Task Force, the Mayor reminded his fellow citizens (or the civic-minded man, at least) that economic regeneration began at home:

> there are 2.5 million people in this city who have one heck of a lot of economic clout. Let's put it to use. Treat your wife or your girlfriend to

dinner out. Stay in a hotel for the weekend. Go and see a show or cheer on the Jays.[44]

Thus, even before medical efforts to eradicate SARS had succeeded, local government and business leaders urged Torontonians to prove that the city was 'safe to live in and to visit', as long as one avoided hospitals. Most importantly, if more residents walked the streets and crowded into restaurants and sporting events foreign journalists might be prompted to forget Toronto's plaguetown identity.

Fueled by a 5 million-dollar budget, an aggressive, multi-pronged marketing campaign to refamiliarize the city began the day after the WHO announcement. Treating the city as a product and its citizens as the prime market, revival tactics turned Torontonians into tourists. Mayor Lastman initiated rebranding efforts under the banner of 'Toronto You Belong Here'. At a press conference geared toward Mother's Day celebrations, the Mayor decoded the catchphrase: 'this simple slogan shows that Torontonians are the warmest, most welcoming people on earth'. The logo conveyed similar sentiments, though in an understated way: the word 'Toronto' appeared in a plain, lower-case font, with 'you belong here' added underneath, in hand writing that evokes the look of a homely note.[45] Advertising firm BBDO Canada Inc. devised the slogan and also assembled a series of print, radio and television advertisements in local media 'to engage Torontonians to enjoy their City'.

Performing one's civic duty by behaving like a tourist is a task far less onerous than adhering to a public health officer's quarantine order, but BBDO and city tourism leaders recognized that fearful residents needed prodding. If Torontonians missed press conferences and advertisements, then they might meet a team of 'young and energetic uniformed ambassadors', who were posted at the main train station and shopping malls. One of the ambassadors' tasks was to distribute 'Toronto: You Belong Here' packages, which included information about city features along with coupons relating to local 'signature' events, such as fireworks displays. Another of their jobs was to greet passengers at the airport, in conjunction with Air Canada's 'Canada Loves Toronto' campaign. When traveling their ambassadorial beats they drove SUV's 'branded with the "Toronto: You Belong Here" slogan'. This un-Canadian orgy of self-love and self-promotion was more than an attempt to compensate for the drop in tourist arrivals: at a deeper level the belonging campaign reassured Torontonians that they need not feel like 'lepers' or 'pariahs' on the global health stage. This was not plaguetown: it was *our* town.[46]

If New York and San Francisco had signature songs, then why not Toronto? BBDO commissioned 'The Toronto Song' to induce civic pride and to reacquaint Torontonians with their home town. The winning entry, 'Right Here with Me', played on local radio stations and provided the audio background for 30 and 60-second commercials, which advertised local festivals and attractions, such as the 'Celebrate Toronto Street Festival' and the Molson Indy car race. A folk-rock tune, performed by Jason Gleed, (a local composer and songwriter) accompanied lyrics about separation and joyful reunion ('cause you belong right here with me', the chorus calls out boisterously). But the song's bridge, introduced with a minor chord and sung at half tempo, expresses the singer's gloominess over abandonment:

> You've gone and done it again
> You forgot where you came from
> But you can't erase your home
> Cause when you run away
> You're only running from yourself
> You don't need anyone else –
> Except me

In the voice of a jilted lover, hurt but confident that his amour will return, the singer expresses the feelings of a city troubled that its own citizens turned their backs on their home town. But after the song's dark bridge, its up-beat, anthem-like chorus powers toward a future of togetherness. As the advertising firm explained, the song and the commercial were selected to 'captur[e] the emotional bond of Torontonians to their city'.

The Toronto Song was not commercially released but it was posted as a downloadable MP3 file on the official City of Toronto website. All of the 'you belong' promotional material and commercials directed visitors to the site, which provided a regularly updated events calendar as well as coupons for local attractions. Another downloadable feature available in mid-2003 was a virtual postcard. By clicking the 'Create your own City of Toronto postcard!' link, a generic postcard template popped up. Unlike the news photos of decimated Chinatown thoroughfares and empty malls that continued to circulate in 2003, the recovery campaign's image of Toronto presented the downtown Toronto skyline from a distance, with the 'you belong' logo on the bottom left. The vantage point is the shore of the Toronto Islands, and the point of view aligns with that of a lone cyclist, pausing to take in the view on a warm day. A sailboat skips by and the city's most recognizable buildings, the CN Tower and the SkyDome stadium, both major tourist attractions, balance financial towers and hotels. On the right sidebar, two city scenes (a fashionable café patio and the streetscape of the city's main theater strip) along with an image of spectators filling the SkyDome accompany a close-up of a cuddly-looking polar bear cub. Yet another bear, a grizzly, adorns the virtual 8-cent stamp. Although neither type of bear lives anywhere near Toronto (save those on display at the Metro Toronto Zoo), they are recognizably Canadian totems that represent the entire country as a barely-tamed wilderness. In concert with the Toronto Island view, the animals distance the viewer from the urbanity of Canada's biggest city, where a mystery disease was still claiming lives. The postcard slogan, again in a memo-style handwriting font, further connotes tentativeness toward the city in the phrase: 'It's time for a little T.O' (the already-established nickname for Toronto). Senders could not rearrange or alter the images but the site offered content options from the 'Select your Postcard's message!' link:

> Having a wonderful time.
> Wish you were here!
> Toronto is the best!

In announcing the launch of the postcard on 17 July, Mayor Lastman declared, 'I want this postcard on bulletin boards from Paris to Beijing. Come on, Toronto, let's show the world how much we love our city'.[47] In spite of its generous sprinkling of exclamation marks, however, the card conveyed ambivalence and reluctance to embrace a city still smarting from SARS: the 'you belong here' logo appeared snugly on the *island* side of Toronto's harbor, and only a 'little T.O.' was recommended.

Toronto: World City

Once quarantine and isolation measures appeared to end the transmission of SARS less apologetic campaigns, directed particularly to US tourists emerged. Harkening back to the City Tourism Commissioner's 2002 report, which had identified 'Toronto the Diverse' as a potential brand, this Tourism Toronto initiative emphasized the city's cosmopolitan character. Just as post-independence African nations and post-colonial creative artists market the margins by catering to Western tastes, so this campaign to restore Toronto's image as a safe and appealing destination strategically exoticized the city, symbolized through its new slogan, 'Toronto – the World in a City'. After having been marked in racial terms as a 'pariah' and designated out-of-bounds for business and leisure travelers, Toronto was repackaged (by the Provincial tourism ministry, the city, and the Toront03 Alliance, the new civic organization, which helped to organize 'SARSstock') as a place where the exotic resided in 'ethnic' restaurants and enclaves, not hospital wards.

The SARSstock mega concert on 31 July 2003 was a one-day spectacle that demonstrated the new boldness in recovery initiatives. Small-scale entrepreneurs (who had sold surgical masks by city roadsides 3 months earlier) switched to selling souvenir 'I survived SARS' and 'SARS-stock' t-shirts at the concert. Half a million spectators, as the *Toronto Star* put it, were 'carriers of good news'; that so many people had mingled without fearing SARS confirmed that 'the city has survived'. Headliners, such as the

Rolling Stones, Justin Timberlake and AC/DC drew world media interest, spinning 'positive stories' in 564 North American newspapers and on several US television networks. Rather than beleaguered white-coated doctors spouting SARS transmission theories and lashing out at the WHO, the city acquired splashier spokespeople. Strutting out on stage in a fuchsia morning coat, Mick Jagger screamed: 'Toronto is back, and it's booming'. Whipping up the crowd, he added: 'This is the biggest party in Toronto's history, right? You're here. We're here'.[48] Likely unaware that his words echoed the 'you belong here' recovery slogan and the 'Right here with me' song, Jagger made the sentiment sexy by embracing the city enthusiastically as an entertainment mecca.[49]

Aside from the concert, the recovery campaign continued to focus on refamiliarization, and reassurances that the city was safe, even if 44 of its citizens had died. In the summer and early fall of 2003, politicians and business leaders became convinced that aggressive marketing was necessary. The product was there, in the opinion of Toronto's first openly gay City Councilor, Kyle Rae. Having just hosted guests in town for the city's Lesbian and Gay Pride festival he noted that their impression of the city was

> one of shock and amazement that Toronto, even after SARS, has a clean healthy downtown, lots of destinations, places to go, nightlife, art galleries, museums, great shopping. We've got everything people are looking for...It's just that we're not marketing it properly.

The president of Molson Inc., the brewery that was the primary private sponsor of the concert agreed: 'We've taken the first step in regaining our image...We've got to keep selling ourselves'.

The second major campaign, sponsored by the Province and Tourism Toronto came with a new, trademarked slogan, 'the world within a city'. The major coordinating component of this drive was and remains the Toronto Convention and Visitors' Bureau website, which incorporated the content and design of a website first designed by the Toront03 Alliance. 'Welcome to the World's Best-Kept Travel Secret', announces the 'Destination Toronto', homepage. Site visitors discover different elements of the secret, one of which is that Toronto is home to 'more than 80 cultures from around the globe'. It also offers 'international' cultural festivals, shopping with 'international flair', and 'vibrant features representing every part of the globe'. Thus, far from cringing at its status as a metropolitan crossroads, in which outsiders might at any moment introduce exotic diseases, the site extols Toronto's connectedness to the world. The 'Personality' page, which deconstructs the city's new slogan, explains precisely how Toronto embraces, but does not absorb diverse cultures:

> *The world within a city™* doesn't just refer to the wealth of dining, shopping and theatre experiences Toronto has to offer...Toronto is known as one of the most culturally diverse cities in the world, and we take pride

> in the knowledge that we do not have to be homogenous to have a peaceful, thriving city.

Moving sharply from the brink of overt anti-Chinese racism, the second phase of the tourism recovery strategy turned to a cherished Canadian myth, that Canada is a tolerant, multicultural country, happily free from race wars and ethnic and religious clashes – and, by implication, free from exotic diseases as well. Unlike the 'You Belong' website, which initially included a link to a separate SARSTORONTO.COM information site, the World Within a City site yields no hits if one searches for 'SARS'.

The worldly campaign engages with the exotic strategically, in an effort to lure tourists and business travelers looking for something interesting but predictable. It invokes the colorful and unusual in romantic, non-threatening terms. References to its 'vibrant and quirky neighborhoods' are codes for ethnic and sexual diversity: 'some [neighborhoods] can be described as "ethnic", while others reflect a particular lifestyle, business, or leisure activity' (what follows is a list: 'Greek town, Little Italy, Chinatown or the Gay Village'). While the visitor might encounter 'delightful secrets at every turn', there is nothing inscrutable about the place, no nasty surprises to encounter when 'neighborhood hopping'. In fact hosting 80 cultures does not make Toronto 'foreign' because of the city's 'amazing ability to adapt and flourish, while remaining essentially 'Canadian' – gracious, broadminded, safe and friendly'. The subtext here is safer and friendlier and more tolerant than the US, and thus a haven for US visitors. For example the 'fact file' page addresses Americans explicitly by mentioning that Toronto is located on the same latitude as northern California: '(remember that the next time it snows in California.)'

According to the World within a City site, Toronto not only has a large Asian population; more intriguingly, the Chinese comprise 'one of the largest and most visible ethnic communities'. Like the local press, which had focused on the first SARS patient's links to Hong Kong, the 'Chinatown' page notes that 'many of the city's Chinese residents are relatively new immigrants from Hong Kong'. In this recovery context, however, the commercially-active Chinese bring an approved and non-threatening exoticism to the city, and any resemblance to China is now a plus, as we see in this description of Chinatown's main intersection: 'It has grown into a frenzied, boisterous neighborhood that can easily be mistaken for Hong Kong'. Even though many medical experts agreed that the transference of the virus from exotic animals to humans in Guangzhou markets had allowed SARS to emerge, the World within a City site extols Chinatown's authentic market atmosphere:

> on weekends – especially, the sidewalks are crammed with open-air food stalls, vendors and thousands of people from all backgrounds eager to shop, eat and socialize.

Chinatown's food is equally authentic, offering 'exotic fare' and a wide selection of teas, 'some unusual to Western tastes'. Thus, even this latest and expensively produced campaign to re-brand Toronto draws on centuries-old notions of exoticism, with the Chinese the chief representatives of 'the unusual'. But equally notable is the lack of tentativeness about the 'vibrant' city. Six months after SARS hit, city tourism promoters determined that it was time for a *lot* of T.O., and a lot more confidence that the city could wrap the exotic safely and securely, Canadian-style.

Conclusion: Toronto, a remarkable brand

Critical studies of public health emphasize that marginalized populations – the poor, the poorly-housed, the indigenous, and stigmatized ethnic groups – have historically borne the greatest burdens of medical policing.[50] And in North America, the Chinese have often been singled out for discriminatory treatment, consistent, as Shah notes, 'with the logic of public health measures that routinely conflated deadly disease with Chinese race and residence'.[51] The outbreak of SARS in Toronto in 2003 replayed these historic themes: a baffling virus, a suspected Asian source, restrictive quarantine. But in this instance media narratives and images (compounded by confused pronouncements from medical and civic authorities), rather than officially-imposed and discriminatory measures, gave Toronto's SARS identity as plaguetown a racial cast. Once medical officials announced that SARS had emerged, Toronto acquired an unprecedented 'publicity problem'. In a matter of days the city underwent a discursive transformation from a technologically-advanced North American city into a dangerous place. Images of empty streets and abandoned malls confirmed that a climate of fear reigned: residents were hiding, visitors beware. Alarmist coverage of the city's viral status unhinged Toronto – a city Peter Ustinov once nicknamed 'New York run by the Swiss' – from its smug, Western identity and hitched it to Asia, turning the erstwhile North American city into an object of Western curiosity.

Since Toronto's tourism industry had been in the doldrums before SARS appeared, tourism promoters saw the SARS crisis as a chance to act boldly to recast Toronto's image.[52] In addition to the 'You Belong' and 'World within a City campaigns', a coalition of agencies announced in April 2004 a new project to re-brand the city. 'Re-imagining Toronto: the Quest for a Remarkable Brand', set the challenge for advertising and marketing firms.[53] The winners, Brand Architecture International (along with local advertising firm TWBA/Toronto), began with a public campaign, called, 'We are Toronto'; its object was to elicit responses from Torontonians, to 'help us discover what makes Toronto different, better, special and distinct from other cities of the world'.[54] The conclusion, however, was already written: ' Toronto is an amazing city and we want the world to know!'[55] Evidence of racial profiling, harassment, job loss, and the media stigmatization of Asians in a 'whitish' city is unlikely to appear on the marketers' lists of things that people ought to know about 'diverse' Toronto.[56] As the Chinese National Council's 'Yellow Peril' report cautions, analyzing recovery from public health scares solely in medical or business terms fails to expose how easily ideals of 'anti-racism and human rights' slip away. While explicit racist stereotypes have faded in Canadian public discourse, they became vivid once viral danger could be linked to an Asian source.[57] Thus, whether or not brand architects acknowledge it, Toronto's exoticization and the orchestration of its image makeover cannot be understood outside of historically-rooted discourses of racial, national, and Western identities, which both constituted the mysteriousness of the virus and informed mediated strategies for the city's renaissance. 'Strong brands are built on a foundation of truth', the Toronto Branding Project declares, but in practice, half-truths will do.

Notes

1 J. Coyle, 'Exactly what we needed', editorial, *Toronto Star* (31 July 2003). http://www.chariotmedia.com/TorontoStar (accessed 1 July 2004). Senator Jerry Grafstein made the connection between media communication and plague transmission.

2 A. Nikiforuk, 'Fever Pitch: SARS is a reminder of the economics of disease', *Canadian Business*, 76, no. 10 (26 May 2003): 119–21.

3 P. Slack, 'Introduction', *Epidemics and Ideas: essays on the historical perception of pestilence* (Cambridge: Cambridge University Press, 1992), pp. 1–20, 12–13; M. Healy, *Fictions of Disease in Early Modern England: bodies, plagues and politics* (London: Palgrave, 2001); A.M. Kraut, *Silent Travellers: Germs, Genes and the 'Immigrant Menace'* (New York: Basic Books, 1994), pp. 24–30.

4 A. Bashford, *Imperial Hygiene: a critical history of colonialism, nationalism and public health* (London: Palgrave, 2004), pp. 117–18; N. Shah, *Contagious Divides: Epidemics and Race in San Francisco's Chinatown* (Berkeley: University of California Press, 2001), pp. 122–3.

5 The Ontario Ministry of Tourism estimated that the loss to the provincial tourism sector was two billion dollars, and that 28,000 jobs were lost. S. Thomas, 'A Year after SARS, tourists return to Toronto', 26 May 2004 (Reuters). www.reuters.com/printerFriendlyPopup.jhtml?type=topNews&storyID=5264582 (accessed 1 July 2004).

6 Lastman mentioned this in a speech at a special meeting of City Council, 24 April 2003. www.city.toronto.on.ca/mel_lastman/speeches/sarsreponse_council_2003.htm (accessed 1 July 2004).

7 D. Gilmour, 'Coming Soon', *Toronto Life*, 38, no. 2 (February 2004): 60–5. Five images, each of the 1917 influenza epidemic, accompanied this story.

8 For example, 'We are home to a higher proportion of immigrants than any other city, surpassing Miami, Sydney, Los Angeles and New York'. Toronto City Summit Alliance, 'Enough Talk: An Action Plan for the Toronto Region' (April 2003): 19.

9 A representative film in this genre is *28 Days Later* (2002). For a SARS-framed review of the film see E. Helmore, 'Plague upon the Brits: Not quite a SARS film, '28 Days Later' depicts viral devastation in London', *New York Daily News* (23 June 2003).

10 Toronto's Chief Medical Officer of Health reflected: 'The SARS outbreak illustrated the vulnerability of Toronto to a new emerging disease and showed how vital a strong public health infrastructure is the well being of the city'. Dr. Sheela V. Basrur to Board of Health, in Toronto, *City of Toronto Staff Report* (9 September 2003): 13.

11 D. Gilmore, 'Coming Soon', *Toronto Life*, 38, no. 2 (February 2004): 60–5. The subtitle of the article reads: 'A year after SARS, here's what we've learned: 2003 was just a dress rehearsal for the inevitable pandemic', p. 60. See also 'Killer pneumonia finally arrives: we were overdue for an epidemic', Toronto *Globe and Mail* (18 March 2003): A17; 'Where's Rudy Giuliani when you need him? Yesterday the World Health Organization clapped the city of Toronto into quarantine. SARS became Toronto's 9/11, the worst crisis in its history'. Toronto *Globe and Mail* (24 April 2003): A1. These were all cited on the 'SARS headlines' page, CBC News Online website: http://www.cbc.c./news/background/sars/headlines.html (accessed 1 July 2004).

12 News reports suggested this 'exotic' link early on in the crisis, but the theory received authoritative medical endorsement in the *New England Journal of Medicine*. See R.P. Wenzel and M.B. Edmond, 'Managing SARS Amidst Uncertainty', *New England Journal of Medicine*, 348 (2003): 1947–1948. The Naylor Report reiterated the connection: 'It now appears that exotic animals in a Guandong market – perhaps civet cats or raccoon dogs – may have given the human race yet another novel infectious disease'. *Learning from SARS: Renewal of Public Health in Canada. Report of the National Advisory Committee on SARS and Public Health* (Ottawa: Government of Canada, 2003), p. 23.

13 *Learning from SARS: The SARS Commission Interim Report: SARS and Public Health in Ontario* (Toronto: Government of Ontario, 2004); *For the public's health: A Plan of Action. Final Report of the Ontario Expert Panel on SARS and Infectious Disease Control*. (Toronto: Ontario Ministry of Health and Long-term Care, April 2004). See also K. Donovan and T. Talaga, 'SARS: The chain of errors, *Toronto Star* (20 September 2003): A1.

14 C. Leung, *Yellow Peril Revisited: Impact of SARS on the Chinese and Southeast Asian Canadian Communities* (Toronto: The Chinese Canadian National Council, 2004), p. 2.

15 P. Mason, *Infelicities: Representations of the Exotic* (Baltimore: Johns Hopkins University Press, 1998), p. 1.

16 D. Manderson, '"Disease, Defilement, Depravity": Towards an Aesthetic Analysis of Health' in L. Marks and M. Worboys (eds), *Migrants, Minorities and Health: historical and contemporary studies* (London: Routledge, 1997), p. 25. On Canada see R. Mawani, '"The Island of the Unclean": Race, Colonialism, and "Chinese Leprosy" in B.C., 1891–24', *Journal of Law, Social Justice and Global Development*, 1 (2003): 1–21.

17 Reuters used the leper reference to describe the effects of the WHO announcement. Quoted in N. Seeman, 'Toronto Under SARS: A Media Fever', *National Review Online* (25 April 2003): 1. The pariah quote was prefaced with: 'Empty shops, quiet streets, hotels with high vacancy rates'. *BBC News online*, 24 April 2003. Cited on CBC News webpage, 'SARS in the headlines', http://www.cbc.news/background/sarsheadlines.html (accessed 1 July 2004).

18 The WHO was equally baffled by the Toronto outbreak: 'Another mystery is why the SARS virus spreads more efficiently in sophisticated hospital settings'. 'Update 93 – Toronto removed from list of areas with recent local transmission', World Health Organization, 2 July 2003. http://www.who.int/csr/don/

2003_07_02/en/print/html (accessed 1 July 2004). Later research confirmed that intubation, less commonly practiced in poorer countries, contributed to the rapid spread of SARS in Toronto hospitals.

19 Five days after the WHO included Toronto in its travel advisory it removed Vietnam from its lists of areas of local transmission. Vietnam had first appeared on this list, along with Toronto, parts of mainland China, Hong Kong, Taiwan, and Singapore, on 22 March 2003, and it was the first 'hot spot' officially to halt transmission. 'Update 92 – Chronology of travel recommendations, areas with local transmission', World Health Organization, 1 July 2003. http://www.who.int/csr/don/2003_07_01/en/print/html (accessed 1 July 2004).

20 In Canada, the mosaic metaphor contrasts the US metaphor of the melting pot. Multiculturalism officially values retaining ethnic and cultural distinctiveness in an immigrant nation. For a critical reading of these declared values see S.H. Razack, ed., *Race, Space and the Law: Unmapping a White Settler Society* (Toronto: Between the Lines, 2002).

21 Sean P. Hier and J. Greenberg, 'Constructing a Discursive Crisis: Risk Problematization and Illegal Chinese in Canada', *Ethnic and Racial Studies*, 25 (2002): 490–513.

22 Leung, *Yellow Peril Revisited*, 7. The Chinese National Council had already been engaged in efforts to force the federal government to provide redress for the Head Tax. See R. Mawani, '"Cleansing the Conscience of the People": Heading Head Tax Redress through Canadian Multiculturalism', *Canadian Journal of Law and Society*, 19 (2004): 127–51.

23 A total of 119 SARS stories appeared in the *Globe* over this period. Leung, *Yellow Peril Revisited*, 10.

24 For example, 'Every case can be linked back to the original index case, a 78-year-old grandmother and immigrant from Hong Kong'. Seeman, 'Toronto Under SARS', 2. http://www.nationalreview.com/script/printpage.asp?ref=/comment/comment-seeman042503 (accessed 1 July 2004). See also, 'How a deadly disease made its way to Canada', Toronto *Globe and Mail* (29 March 2003): A1, A15.

25 Basrur to Board of Health, 12.

26 Gilmore, *Coming Soon*, 64.

27 Helen Branswell, 'Chorus of outraged voices denounces WHO warning against travel to Toronto', Canadian Press Newswire, 25 April 2003, p. 3. See also 'Toronto angry over SARS warning' http://news.bbc.co.uk/1/hi/world/americas/2971217.stm (accessed 13 April 2006).

28 André Picard and Caroline Alphonso, 'Shock, Anger greets WHO advisory' Toronto *Globe and Mail*, 25 April: A1.

29 Kevin Ward, 'British editorial opinion split over Toronto's addition to SARS travel list', Canadian Press Newswire, 24 April 2003, p. 1. A *Times* of London editorial expressed the same assessment: '"The situation [in Toronto] is not at all the same as in Beijing"' (quoted in *ibid*.).

30 'Beijing closes public places over SARS', http://www.cnn.com/2003/HEALTH/04/27/sars.wrap/index.html (accessed 1 July 2004).

31 T. Corcoran, 'SARS a test run for more power', *National Post*, 26 April 2004, p. 11.

32 'SARS according to Mayor Mel', Toronto *Globe and Mail* (26 April 2003): A8.

33 For a cross-aesthetic overview of the exotic's appeal, see D. Lach, *Asia in the eyes of Europe: Sixteenth through Eighteenth Centuries* (Chicago: The University of Chicago Library, 1991).

34 E. Hansen and B. Gordon, *Exotic Destinations* (San Francisco: Full Data Limited, 1994).

35 The London-based *Satiric Press* lampooned the way that SARS publicity brought Toronto world recognition: 'Lastman believes that Toronto will be more respected by the international community now that it's made the WHO's list of most dangerous cities to visit'. http://www.satiricpress.com/sp/archive/archive_articles_bysubject.asp#canadianpolitics (accessed 1 July 2004).

36 G. Huggan, *The Postcolonial Exotic: Marketing the Margins* (London: Routledge, 2001). Huggan argues that post-colonial literature, as well as the academic field of post-colonial studies are both discourses of the exotic.

37 E. Fürsich and M. Robins, 'Africa.com: The Self-Representation of Sub-Saharan Nations on the World Wide Web', *Critical Studies in Media Communication*, 19 (2002): 198.

38 'Increasing Toronto's Profile Internationally and at Home (All Wards)', Report No. 10, Economic Development and Parks Committee, Toronto City Council Minutes, 26, 27, 28 November 2002, pp. 4, 5.

39 'Reviving Tourism in Toronto', Toronto City Summit Alliance, 'Enough Talk, an Action Plan for the Toronto Region, ' April 2003, p. 10. The latter study was conducted over the second half of 2002. It was presented to City Council in April 2003, at the peak of the first outbreak.

40 J. Willis and S.J. Page (ed.), *Managing Tourist Health and Safety in the New Millennium* (London: Pergamon, 2003).

41 'Increasing Toronto's Profile'.

42 T. Noah, 'The Canadian Menace: SARS, Toronto and Danger', *Slate*, 24 April 2003. http://slate.msn.com/id/2081979/ (accessed 13 April 2006).

43 'Fear hurts Toronto as much as SARS: mayor', http://www.cbc.ca/story/news/national/2003/04/24/sars_toronto030424.html (accessed 13 April 2006). See also Jennifer Lewington, 'Lastman's Gaffes Hurt Toronto', *Toronto Globe and Mail*, (26 April 2003): A1.

44 'Remarks by Toronto Mayor Mel Lastman at a special meeting of Toronto City Council', 24 April 2003. www.city.toronto.on.ca/mel_lastman/speeches/sarsreponse_council_2003.htm (accessed 1 July 2004).

45 The words, 'You Belong Here' were scripted by one of the advertising designers, not printed in a professional font. Carolyn Strange interview with Duncan Ross, Executive Director, Tourism Toronto, Toronto, 5 October 2004. Ross commented: 'she happened to have nice handwriting'.

46 G. Galloway and W. Immen, 'Canadians get tough assignment: Love Toronto', Toronto *Globe and Mail* (29 April 2003): A1.

47 http://wx.toronto.ca/inter/it/newsrel.nsf/0/e7c57d31822c704085256df60045c930?OpenDocument (accessed 13 April 2006).

48 'Hot, Sweaty SARSstock a Smash Hit', was the CBS news headline. http://cbsnews.com/stories/2003/07/31/entertainment (accessed 1 July 2004). A Toronto cab driver offered his own media analysis: 'they beamed pictures of the concert around the world, and showed that hundreds of thousands of people can gather, without a SARS mask in the crowd, without any ill effects'. N. Van Rijn, 'City on the road to recovery', Toronto *Star* (2 August 2003): A1.

49 Canadian politicians had more authority but less panache: 'I think it's important we get the message out that Canada is safe', the Deputy Prime Minister stated in a post-concert interview.

50 Examples in this large literature include V. Prashad, 'Native Dirt/Imperial Ordure: The Cholera of 1832 and Morbid Resolutions of Modernity', *Journal of Historical Sociology*, 7 (1994): 243–60; W. Anderson, 'Excremental Colonialism:

Public Health, and the Poetics of Pollution' in A. Bashford and C. Hooker (eds) *Contagion: historical and cultural studies*, (London: Routledge, 2001), pp. 76–105; D. Arnold, *Colonizing the Body: State, Medicine and Epidemic Disease in Nineteenth-Century India* (Berkeley: University of California Press, 1993); M. Swanson, 'The Asiatic Menace: creating segregation in Durban, 1870–1900', *International Journal of African Historical Studies*, 16 (1983): 401–21; Bashford, *Imperial Hygiene*, 1–7.

51 Shah, *Contagious Divides*, 120.

52 D. Seguin, 'Hawking Hogtown: Toronto's Image Problem', *Canadian Business*, 76, no. 10 (26 May 2003): 112–17.

53 The project is a joint initiative of the City of Toronto, Tourism Toronto, the Ontario Ministry of Tourism and Recreation and the Toront03 Alliance. It was initially recommended by City Council in November 2002. See http://www.toronto.ca/branding/pdf/2002_Branding_StaffReport.pdf (accessed 13 April 2006).

54 The Branding Project received 4,500 submissions which 'really helped get a better picture of Toronto today, as well as an excellent idea of what the public wants the city to become'. The team also conducted focus groups in Canada, the US and the U.K. http://www.toronto.ca/unlimited/background.htm#final (accessed 13 April 2006).

55 The 'quote of the month' for April 2004 on the Tourism Toronto website is: 'Diversity not only rules [in Toronto], it's celebrated with a vengeance'. *Carla Waldemar, Lavendar Magazine*: http://torontotourism.com/media (accessed 3 June 2004).

56 Anti-racist and labor groups consistently attempted to draw attention to the racist dimensions of the SARS crisis, and complained that cash infusions were directed to major businesses, rather than the most vulnerable workers. See, Toronto Labor Council, '"SARS": Toronto responds to a major community crisis', http://www.labourcouncil.ca/sars.html (accessed 1 July 2004), and T. Goossen, C. Pay and A. Go, 'Healing the scars in post-SARS Toronto', Toronto *Star* (19 May 2003), http://www.buzzardpress.com/acla/sars/healing_scars_post-sars.html (accessed 13 April 2006).

57 Hier and Greenberg, 'Constructing a Discursive Crisis', 508.

13 The Geopolitics of Global Public Health Surveillance in the Twenty-First Century

Lorna Weir and Eric Mykhalovskiy

Social scientists, historians, infectious disease experts, and others generally recognize the period from the mid-1980s to the present as one of dramatic change in the global relations for knowing and responding to international infectious disease threats.[1] Some commentators have gone so far as to suggest that in the wake of the 2003 SARS outbreak, a near revolution is underway in the global governance of infectious diseases.[2]

Declarations of novelty can have their own infectious quality. Fortunately, historical argument has helped temper the debate about what is precisely new about the emerging field of global public health governance.[3] Discussions typically recognize the origins of established features of current international infectious disease monitoring and surveillance – the International Health Regulations, the organizing principle of minimal disruption to international travel and commerce, the WHO (World Health Organization) structure, and so on – in the activities, emerging institutions, and framework of agreements arising out of the series of International Sanitary Conferences held between 1851 and 1903.[4] In identifying novelty against this backdrop, much attention has been focused on transnational health risks that accompany the processes known as 'globalization', together with new anxieties about emerging infectious diseases, and the increasing political significance of non-state actors such as NGOs, the private sector and social movement organizations in global health.[5]

In this chapter we seek to characterize the specificity of current changes in global health governance through an analysis of new governmental techniques in global public health surveillance. Configured around the virtual technology of the Internet, mobilizing institutional networks centered around the WHO, and drawing fundamentally on health news as an information source, novel techniques of global surveillance are reconfiguring the relation between political borders and knowledge in global public health. Yet they have received little serious attention in the social science and historical literatures.

Our analysis is empirically focused on the history and current operations of GPHIN – the Global Public Health Intelligence Network – an early warning system for global public health events developed by Health Canada in collaboration with the WHO. Unlike prior approaches to global infectious disease surveillance, GPHIN systematically gathers, classifies, translates and distributes online news information rather than epidemiological case reports. It thus provides an important vantage point from which to explore the implications for global public health of shifts in the established boundaries of surveillance knowledge.

We argue that GPHIN constitutes a transformation in the social organization of knowledge of international infectious disease outbreaks. GPHIN is a knowledge technique with multiple effects. GPHIN has resulted in enhanced knowledge, action capacity, and authority for the WHO in the context of global infectious disease control.[6]

GPHIN is an information source beyond national control that gives the WHO the ability to have authorized knowledge of outbreaks globally and to respond to outbreaks while they are occurring. As an electronic early warning alert technique, GPHIN both speeds up the knowledge of outbreak and bypasses its previous routing through sovereign nations. Temporally, GPHIN enabled the WHO and other organizations to have knowledge of outbreak while outbreak is taking place. This temporal capacity has strengthened the WHO in global health governance. And geopolitically, prior to the existence of GPHIN, sovereign nations controlled what information they would share with the WHO in accordance with international agreements. The official country notifications received by the WHO were often years out-of-date and useful essentially for statistical purposes – and even then outbreaks went unreported. The simultaneity of outbreak with international knowledge of outbreak laid the basis for extending the action capacity of the WHO, for instance in sending response teams to sites of outbreak. Online early warning outbreak detection has thus inserted a knowledge technique into the international control of infectious diseases that has precipitated a new social organization of global health governance.

We begin our discussion by specifying the sense in which we understand the term 'surveillance', contextualizing GPHIN in the history of public health surveillance in the second half of the twentieth century. The discussion then moves to tracing the discursive preconditions for online global public health surveillance in the novel concept 'emerging infectious diseases'. We describe how the term was popularized in North America in the early 1990s through the work of the US Institute of Medicine, inciting official and popular concern about 'microbial threats' to human health. We also suggest how the role of the WHO in the current institutional configuration of global infectious disease surveillance came about as a response to the perceived need for global leadership that surfaced when emerging infectious diseases were problematized in relation to the 'spectre' of international microbial traffic.

We then turn to the development and operation of GPHIN based on interviews conducted with GPHIN staff and managers.[7] We show that the initial concept for GPHIN was sparked by events surrounding the 1994 outbreak of pneumonic plague in Surat, India. In particular, we argue that the GPHIN concept arose as an attempt to recuperate international public health in the face of media expectations that it had the capacity to manage outbreak anywhere. Pressure on the WHO to change its way of managing infectious diseases thus had two conditions: 'emerging infectious diseases' and global health news.

We further describe how GPHIN's turn to online health news gave rise to two new related surveillance techniques: early warning outbreak detection and outbreak verification. We argue that their incorporation within the apparatus of global infectious disease surveillance has resulted in a 'post-Westphalian' sourcing and organization of knowledge and action about international infectious diseases.[8] Previously, nation-states held authority over public health information within their borders, sharing information in relation to international treaties and agreements. Currently new distributions of power are being constituted, including links between global public health networks and local authorities, which transgress state boundaries of control over outbreak notification. In our concluding discussion we analyze the integration of global public health surveillance with anti-terrorism initiatives in the Global North, cautioning that this represents a neocolonial agenda that must be questioned for the sake of constituting public health as a global public good.

Problematizing emerging infectious diseases

'Surveillance' is a word freighted with many meanings in contemporary social scientific and historical writing. In this chapter, when we refer to GPHIN as a technique of 'public health surveillance', we follow the meaning of 'surveillance' found in our sources. 'Public health surveillance' was constituted as a field in national public health systems during the second half of the twentieth century. It was conventionally recognized to comprise four kinds of action: collecting, interpreting, disseminating and acting on health information by authorities.[9] Surprisingly, the public health meaning of 'surveillance' as pertaining to continuous data collection about diseases on a population basis is relatively recent, dating to the mid-1950s. Of course, international cooperation in monitoring populations for infectious disease is not new, having been called for at the 1896 International Sanitary Conference.[10] However the term 'surveillance' in public health was applied to individuals, not diseases, in the first half of the twentieth century. Until 1955, 'surveillance' meant closely observing the contacts of a person who had an infectious disease such as plague, smallpox, typhus and syphilis. Contacts were followed to detect the first signs of infection, but without restricting their freedom of movement, until such time as they became symptomatic, at which point they were placed in isolation.[11]

Our present meaning of surveillance as the gathering of population data about disease dates to the studies undertaken in 1955 by Alexander Langmuir at the Centers for Disease Control and Prevention (CDC) when a batch of polio vaccine became contaminated and began infecting children who had been inoculated with it. The collection and analysis of data about the polio outbreak traced the cases to a single source, with the result that the vaccination campaign continued after the contaminated batch had been destroyed.[12] Langmuir, the chief epidemiologist at the CDC during the 1950s and 1960s, is credited with reshaping the meaning of surveillance in public health by applying it to populations rather than individuals, separating surveillance from disease control, extending surveillance to non-infectious diseases, and tying surveillance data to action through dissemination and tactical alliances with government.[13] National surveillance systems for infectious and non-infectious diseases in North America derive from Langmuir's conceptualization of disease surveillance, although the formation of national surveillance systems occurred slowly and fitfully in the second half of the twentieth century. Given that infectious diseases were thought to be essentially solved in industrialized countries from the 1950s until the 1990s, the growth of national systems often occurred in the crucible of disease outbreaks.[14]

Most discussions of the formation of a specifically *global* form of public health surveillance in our present privilege the formative role played by microbial disrespect of national borders in a globalizing era.[15] The effects of globalization, we are told, speed up microbial traffic and heighten the need for disease monitoring at a supranational level. As the SARS epidemic of 2002–2003 demonstrated, and as other chapters in this book analyze, dramatic increases in the scale and pace of international travel can lead to cross-border movement of infectious disease agents without detection or forewarning. The globalization of the food supply, rapid deforestation, and large scale population movement during times of conflict further create conditions for the emergence and spread of infectious diseases across national borders, pointing to the need for strengthened disease surveillance at the global level.[16]

The development of GPHIN sheds light on other dimensions of the discursive and institutional preconditions of global public health surveillance. These relate primarily to the rise of global health news and its challenge to epidemiologically-based forms of official country notification of outbreak. But such health news and the forms of surveillance it feeds into require a robust object. The efforts of a group of virologists, public health officials and other infectious disease experts, working in the US in the early 1990s, fashioned that object through the conceptual innovation, 'emerging infectious diseases'.

In the Global North, the problem of infectious disease generally commanded uneven public attention in the post-World War II period. This was not the case in the Global South, where bacterial, protozoal, viral and other infectious diseases have been consistently regarded as the leading cause of death across all age groups. By the end of the 1950s in the advanced capitalist world the problem of infectious disease was understood to have been more or less solved. This posed a challenge for ambitious infectious disease experts, requiring them to problematize infectious disease and encourage public concern about the threat posed by microbes to human health.

One of the most successful of such efforts was undertaken in the late 1980s and early 1990s in the United States. In 1989, under the leadership of Stephen Morse and with the assistance of Nobel Laureate Joshua Lederberg, a National Institute of Allergy and Infectious Diseases/National Institutes of Health conference on emerging viruses was convened. The conference brought together a group of infectious disease specialists to review evidence on the threats to human health posed by pathogenic microbes. The questions raised by the 1989 meeting were further addressed by a special panel organized by the Institute of Medicine, which in 1992 released the report *Emerging Infections: Microbial Threats to Health in the United States*.[17]

Emerging Infections marshals all of the characteristic arguments used to create concern about infectious diseases: overconfidence in medical progress, public health breakdown, microbial adaptation, and public complacency, all punctuated by the recurring trope of humanity's endless war with its unseen and relentless enemy, the microbe. But the Institute of Medicine report added something new. Drawing largely on the growing HIV epidemic, but also on the increased incidence, in the United States, of such infectious diseases as multidrug resistant TB and Lyme Disease, it popularized a new concept – emerging infectious disease – around which concerns about threats to human health could be focused.

The Report defines emerging infectious diseases as 'clinically distinct conditions whose incidence in humans has increased ... in the United States within the past two decades'.[18] The new disease concept is a bountiful one, including within its terms infectious diseases affecting increasing numbers of people yearly (e.g. Lyme Disease), known diseases whose etiology is now understood to likely result from microbial infection, so-called 're-emerging' diseases which are well-known but whose incidence is escalating (e.g. TB), and diseases resulting from the introduction of new and existing infectious agents from other parts of the world (e.g. HIV, dengue, malaria).

As deployed in the Report, the new concept of emerging infectious diseases produces infectious disease as a current and future problem space. It mobilizes concerns about the 'reality' of infectious disease threats, working against 'complacency' by underscoring the range of diseases, new and established, whose incidence is increasing within US territorial boundaries.

The more specific innovation introduced by *Emerging Infections* is its classification of infectious diseases not by agent – viral, bacterial and so on – but by the social and other conditions related to their emergence. Nearly half of the report is committed to a detailed analysis of the factors contributing to the emergence of infectious diseases. These are explored in terms of an ecological discourse that foregrounds how the fate of humans and microbes are bound together in complex cycles of interaction occurring at such overlapping sites as economic development and land use, technology and industry, international commerce and travel, public health collapse and human population change and behavior.

The combined force of positioning human activity as fully implicated in microbial emergence and the very notion of emergence, which constitutes infectious disease as an active or soon to be active presence, create a dynamism around infectious disease. The concept of emerging infectious disease relocates pathogenic microbes out of a history of long-conquered infections and into a present of persistent and always possible threats to human health, all the while constituting that present as open to human intervention.

The Institute of Medicine Report is a document committed to preserving the health of the nation. Its problematization of emerging infectious diseases as a serious US health threat is aided by representing developing countries as sources of infectious diseases and agents that easily travel across national borders. As a text written primarily for a domestic audience, its chief response to the new problem space of emerging infectious diseases is critique and rehabilitation of domestic public health surveillance and readiness. However, the Report repeatedly inscribes the specter of international microbial traffic and, as such, makes a strong call for coordinating infectious disease control at the global level.

One of the interesting features of the Report's discussion of global disease surveillance is the problem of jurisdiction and leadership. Global disease surveillance was in the very early stages of conceptualization at the time *Emerging Infections* was written. Its components, modes of communication, knowledge bases and lines of accountability were not fully delineated. Global surveillance was certainly not the responsibility of the US government, the Report's main addressee, nor did it easily fit within the established areas of responsibility of a given institutional authority. The WHO, the most obvious potential organization to lead an initiative to establish global infectious disease surveillance, is represented in *Emerging Infections* with ambivalence. The Report both lauds the WHO for its work on smallpox and criticizes it for lacking a mechanism to enforce the terms of the International Health Regulations.

Emerging Infections helped open up discussion about the WHO's potential leadership role in global disease surveillance. It specifically recommended that US representatives to the World Health Assembly bring the discussion forward at the WHO. A follow-up Report issued by the CDC, *Addressing emerging infectious disease threats: A prevention strategy for the United States*, put further pressure on the WHO to strengthen its global surveillance activities. The CDC report suggested that a global consortium of research centers for detecting, monitoring and investigating emerging infections be established, 'under the direction of an international steering committee, possibly chaired by the WHO'.[19]

The ambivalence towards the WHO found in both the Institute of Medicine and CDC Reports reflects broader concerns about its role in global surveillance that emerged as part of the problematization of emerging infectious diseases. Garrett's journalistic account of the period suggests the many ways that the WHO was found wanting. The WHO's capacity to intervene in member states to control local outbreaks was severely curtailed by the requirement to respect national sovereignty. The Geneva head office was also regarded as invariably at odds with poorly-staffed regional offices.[20] In addition, the WHO lacked an emergency response office for epidemics, which delayed the entry of international teams into the field and left them uncoordinated.[21] WHO capacity to engage in infectious disease surveillance was thus framed as needing urgent change, with mounting pressures placed on the WHO to increase its surveillance and response capacity to make the control of emerging infections an organizational priority.

By the mid-1990s the WHO had begun to act. Most significantly, in response to the concerns raised in the Institute of Medicine report, the WHO recognized 'emerging infectious diseases as worldwide problems requiring global leadership'.[22] In 1994 and 1995 two international meetings were held in Geneva to define the nature of the WHO's contribution to the challenge of emerging infectious diseases. By 1996 the WHO had developed a strategic plan for strengthening its activities in global disease surveillance based on a mandate it had received from the World Health Assembly in 1995. The WHO strategy adopted the main recommendations of the CDC Report and led to an expansion of laboratory, training and other WHO-coordinated global networks engaged in surveillance, that were linked primarily by the Internet.

In addition, in 1994, the WHO established the Division of Emerging and other Communicable Diseases Surveillance and Control[23] to improve national surveillance systems and coordinate relations with NGOs, expert advisors and collaborating centers. In 1997, the WHO initiated the formation of the Global Outbreak Alert and Response Network (GOARN) which was formally launched in 2000.[24] GOARN links institutions and existing networks to pool human and technical resources and coordinate responses to outbreaks considered to have international public health significance.[25] It currently coordinates field teams for more than 50 outbreaks per year and has 120 institutional partners worldwide.[26]

GPHIN and the changing geopolitical borders of infectious disease surveillance

At roughly the same time that infectious diseases specialists at the CDC and elsewhere in the United States were debating the potential leadership and parameters of a global system of infectious diseases surveillance, a group of enterprising public health officials from Canada was developing its own vision of global public health surveillance. Our analysis of interviews with the founders and current staff and managers of GPHIN suggests how the initiative arose in response to shifts in the conditions for knowing infectious diseases of potential international significance. More specifically, the initial GPHIN concept was invented in relation to overlapping concerns about dramatic increases in the scale and pace of international travel, the limitations of the established WHO-coordinated system of official country notification of outbreak, and challenges posed to official notification by the widespread emergence of online global health news.

Like their US counterparts, the originators of GPHIN were concerned about the potential heightened spread of infectious disease across territorial boundaries enabled by globalization. Unlike the leadership of the infectious disease community in the US, however, the solution they formulated transgressed the established knowledge relations of infectious disease surveillance.

The developing vision of global surveillance advanced by the Institute of Medicine and CDC reports as well as the WHO strategic plan was firmly rooted in the conventional knowledge sources of existing national programs of disease surveillance. It was based fundamentally in the epidemiological case report and called for an expansion of international networks of diagnostic and laboratory infrastructure, trained personnel, and revamped linkages with academic and research institutes to facilitate the identification of disease agents, case definition, and the conduct and dissemination of epidemiological field research.

For the originators of GPHIN, by contrast, global public health surveillance would need to be decoupled from an exclusive reliance on traditional epidemiological methods. One member of GPHIN expressed this sentiment, emphasizing how quarantine measures and the use of established epidemiological surveillance had been rendered all but redundant under conditions of contemporary air travel. Under earlier circumstances of cross-Atlantic travel by ship, the identification of outbreak and the application of preventative quarantine measures were typically signaled by the appearance of disease symptoms. Ship-based travel provided plenty of time for passengers with infectious diseases to become sick and show visible signs of illness.[27] With the time compression of air travel, however, people with infectious diseases cross international borders in a matter of hours, often before the clinical manifestation of disease. When travel cycles occur well within infectious disease incubation periods, outbreak investigation and surveillance can no longer rely on traditional epidemiological investigation. What was required were approaches that would alert public health authorities to the prospect of outbreak more quickly than was possible through epidemiological confirmation of clinical cases.

The understanding shared by the founders of GPHIN that global infectious disease surveillance would require a substantial improvement in the speed with which outbreaks became known was further supported by their critique of the system of international surveillance occurring under the auspices of the WHO. One might describe that system as a Westphalian one.[28] Under its terms, sovereign nations controlled what information they would share with the WHO in accordance with the international agreements into which they entered. International news of outbreak and epidemic was regulated primarily by the International Health Regulations which required signatories to report, within a 24-hour period, outbreaks for only three diseases: plague, yellow fever and cholera.[29]

In contemporary public health, the surveillance system that existed at the WHO from the 1940s to the mid-1990s has come to be called a 'passive' system. One of our research participants emphasized the limitations of passive surveillance due to its reliance on state sources:

> I should back up and say that the system that was in place for years and decades is a kind of passive surveillance system. The local area in a country is expected to report to some kind of regional area health district and they're supposed to report up to the national government and then the national government, if they so elect, will contact WHO... The WHO had to be very passive about that for political reasons. Although offline there would be insistence that the country please report, so that WHO could say something about it, that could be on relatively frequent occasions ignored by the countries.[30]

WHO's system of official country notification was beset by problems of delay, incompleteness and occasional concealment on the part of member states due to weak national public health surveillance systems and the fear of the often enormous economic and political consequences of revealing information about local outbreaks.[31] Twenty-four hour country notification of the WHO was required solely for yellow fever, cholera and plague, leaving out other communicable diseases, notably the emerging infectious diseases of concern to the Global North. WHO responses to local outbreaks were organized through focal point groups responding to the three internationally reportable infectious diseases. Non-official local information as well as specimens would also be sent to the WHO and its collaborating centers by NGOs such as Médecins *Sans Frontières*. Country inquiries were sometimes initiated through WHO diplomatic channels. Older 'passive surveillance' had active elements, yet outbreaks were often reported months, even years, after an event had occurred or were included as part of annual statistics. As another research participant suggested of the latter practice, outbreak reports made 'behind the curve' might have held some value for future prevention work but did nothing to help respond to the reported outbreak.

For the originators of GPHIN, an alternative to the cumbersome, limited, and delayed system of extant international surveillance was first suggested through their experience of the 1994 outbreak of bubonic plague in Surat, India. The Surat outbreak received widespread media coverage. CNN, set up in 1985 to broadcast global news 24/7,[32] covered the epidemic, releasing footage of people fleeing the city. The CNN coverage fed widespread public fear of plague, generating the need for public health officials to monitor events:

> And we were watching that [CNN coverage – EM & LW] and of course that immediately gave rise to some anxiety in Canada because we have such a large Indian community and we have so much travel between the two countries. Even to the point where...we recognized that there was already a flight en route to Toronto by Air India and it threatened work stoppage at the Pearson Airport because this plane might be hauling plague... So...we tried to monitor this thing.[33]

The efforts of Canadian public health officials to monitor the bubonic plague epidemic in Surat were hampered by poor international surveillance information. The federal government of India refused to publicly acknowledge the outbreak despite the CNN coverage[34] and delayed its official report to the WHO.[35] The WHO, consequently, was initially shut out of local events in Surat. Through their experiences with the 1994 outbreak, the originators of GPHIN came to recognize how their own monitoring efforts were being triggered not by international surveillance information, but by international news coverage. As one research participant put it 'the media was so far ahead of the health sector in monitoring and reporting...that the media was driving the reaction, not the World Health Organization'.[36]

In their ongoing efforts to monitor the situation in Surat, the GPHIN founders made online contact with a physician based there. The resulting email exchange brought into relief possibilities for exchanging information about outbreak through electronic media that could outpace official country reports:

> We started to have an Internet conversation with him and he'd tell us every day how many new patients had been admitted to hospital. And

> all of a sudden it dawned on myself and another gentleman here...that we had information the government and media didn't have. And it was all informal. And it occurred to us that maybe the world was changing and that maybe there were other ways to get information besides waiting for the government.[37]

The events in Surat were a pivotal moment in the formulation of GPHIN as a contribution to global infectious disease surveillance. They clearly called into question the authority of international public health authorities to communicate with the media, and to monitor and report outbreak. An alternative source and form of information – global health news – had stolen the initiative. Not only had global health news outpaced official public health reporting, it had driven public response to the outbreak.

While the epidemic in Surat had shown that the WHO lacked basic knowledge of outbreak worldwide and had weak relations with an increasingly global media, it also suggested to those who developed GPHIN the possibilities of using 'non-traditional' forms of information for global surveillance of infectious disease. The proliferation of the Internet and global health news could be turned into an ally if public health authorities were to use it as a source of information, creating the possibility for a new technique of global infectious disease surveillance – early warning outbreak detection.[38] Actively pursuing media reports of potential public health events could overcome the limitation of passive surveillance and help close the gap between the moment of outbreak and the time of reporting. The new surveillance system harnessed new capabilities by interiorizing global health news as its information source.

When, in 1996, the Canadian federal Treasury Board announced a competition among federal departments to encourage uses of the Internet, the public health officials who had been involved in the Canada-Surat email exchange in 1994 seized the opportunity. The proposal to Treasury Board was successful and work began on the GPHIN 1 prototype in 1997, with the system becoming operational in 1998. The GPHIN concept was formulated at that time as an electronic monitoring tool for early warning outbreak detection based on online news sources. The concept was distinct from a related electronic initiative, ProMED-mail, that had already broken with reliance on epidemiological reports for global outbreak detection.

Established in 1994, ProMED-mail is a medically-moderated chat line that brings together and distributes a mix of news reports, online summaries, official and local observer reports, and subscriber information in order to hasten the process by which public health authorities in the USA and internationally receive reports of outbreak.[39] GPHIN, by contrast, is fundamentally sourced in news. The information it scans is publicly available, does not rely on member postings, and is not subject to the processing delays involved in editorial monitoring of a chat line.

GPHIN is perhaps best described as a secure Internet-based global monitoring system for outbreak alert. GPHIN collects information on disease outbreaks and related public health events by monitoring global media on a 24-hour-a-day 7-days-a-week basis. GPHIN's main sources of outbreak information are the online global news aggregators Factiva and Al Bawaba. GPHIN uses an automated scanning system with a custom-built taxonomy of key words and Boolean search syntaxes to identify news items of potential relevance. The GPHIN system currently scans six key areas for news of global public health events: infectious diseases, biologics, and chemical, environmental, radioactive and natural disasters. The automated processes of scanning, filtering for relevancy and categorizing information are evaluated and supplemented by the work of GPHIN's six analysts. In November 2004, the multilingual GPHIN II was launched. Through the use of an automated translation technology developed through a collaborative research agreement between GPHIN and Nstein technologies, GPHIN now functions in all official language of the United Nations. On any given day GPHIN retrieves approximately 2,000 to 3,000 news items of which roughly one-quarter to one-third are discarded as duplicative or irrelevant. The remaining items are sorted by GPHIN analysts and posted on GPHIN's secure website for use by its over 100 users nationally and internationally.[40]

GPHIN is sometimes disparaged as a monitoring device designed for intervention, lacking a basis in case reports, and devoid of analysis, rather than a true surveillance tool rooted in the rich soil of clinical and laboratory diagnoses.[41] It is said that using news as a source is less reliable than case reports, the benchmark of public health surveillance. Yet news is more accurate than is often supposed. For example, an analysis of a 7-month period of ProMED mail reports indicated that 1.7 per cent of official outbreak reports were retracted because of inaccuracy compared with only a slightly greater 2.6 per cent of media reports.[42] An internal GPHIN study showed that it had a 95 per cent accuracy rate.[43]

The news reports upon which GPHIN is based are unofficial sources of information that do not fall within the warrant of governmental or scientific truth. GPHIN developers thus faced the problem of devising a mechanism of authorizing unverified news reports in order for the system to be of broad public health relevance. Their answer took the form of a new application for a concept that had previously been used during the WHO's smallpox eradication campaign:[44] outbreak verification. In an interview, one of the founders of GPHIN emphasized that the initiative could not be sourced fully within Canada. Canada, or any nation-state for that matter, lacked the capacity to provide the kind of verification that was needed. In his words,

> You could not call the Minister of Thailand, or wherever, as the director of GPHIN and say 'we've been getting these reports of outbreaks in Thailand'.

> The Minister could turn and say, 'Well, we have no responsibility to do anything or to respond to your call in any particular way'.[45]

The only possible organization to provide the required verification was the WHO since it alone had the international diplomatic mandate to make inquiries of its national member states through the country representatives of the World Health Assembly. After a two-year period of collaboration and implementation begun in 1999, the WHO and Health Canada, on behalf of GPHIN, entered into an agreement that GPHIN would supply the WHO with monitoring data and the WHO would verify the reports through its official country contacts.

The combination of GPHIN's online early warning outbreak detection and the WHO's verification capacity provided an effective response to the global health media's challenge to the credibility of international public health authorities. The collaboration created a means of bypassing what had become an ineffectual system of international outbreak notification based on official country reports, speeding up public health knowledge of outbreak so that it coincided with the time of outbreak. While infectious disease surveillance had since the 1950s drawn on news sources,[46] GPHIN's exclusive sourcing of surveillance in news is without precedent.

At the same time, GPHIN has weakened national control over the announcement of outbreak, constructing a global space with national relays. The collaboration between GPHIN and the WHO created an alternative to the state-controlled 'pyramid' of epidemiological reporting from local to regional to national and international health authorities. As a member of GPHIN observed: 'We were squashing the pyramid down to a flat plain in which information could come from any particular place at any time. And governments were no longer in control of their information'.[47] The GPHIN-WHO collaboration unsettles the borders of state secrecy and control over public health information. At the present time, sovereign states are no longer able to contain news of outbreak within their borders, as the People's Republic of China came to understand during the SARS outbreak, which GPHIN first called to the attention of WHO in November 2002. Of course, sovereign control over domestic public health information has not been fully eclipsed. GPHIN's technological innovation has been responded to by China, for example, in the form of censoring domestic internet information.[48]

The move from 'pyramid' to 'flat plain' further marked a shift in the action capacity from launching infrequent responses to outbreak to the potential for intervention at the time of outbreak, a profound change in global response capacity. The collaboration between GPHIN's online early warning outbreak alert capacity and the WHO's outbreak verification capacity created the conditions for the emergence of the WHO as the pre-eminent authority in global public health surveillance, albeit with national rivals.

The securitization of global public health

Ilona Kickbusch, one of the leading commentators in the emerging field of global public health governance, has argued that the increasing focus on the narrow issue of infectious diseases within global public health has been accompanied by a move, particularly in US policy, to distance development health aid from its former humanitarian rationale. Funding to global health is increasingly viewed through the lens of national economic, political and security interests. Within the field of public health surveillance measures, security concerns predominate, fueled by the experience of the intentional spread of anthrax in the USA after the destruction of the World Trade Center.[49]

Although there was increasing integration of public heath surveillance with US national defense after the events of 9/11, the concept of 'emerging infectious diseases' had from the first been formulated in relation to US national interests. As mentioned earlier, the watershed report from the National Institute of Medicine in Washington, DC, had tellingly been titled, *Emerging Infections: Microbial Threats to the United States*[50] and was followed by the CDC's 1994 report, *Addressing Emerging Disease Threats: A Prevention Strategy for the United States*.[51] A decade prior to 9/11, then, 'emerging infectious diseases' were spatialized in geopolitical terms as 'threats' to the 'American nation'. ProMED-Mail was established to detect and protect the USA against the intentional spread of pathogens.[52] The interdigitation of 'emerging infectious diseases' and bioterrorism continued in the mid-1990s, when, in the aftermath of the collapse of the Soviet Union, the extent of the Soviet biological weapons program first became known.[53] Western powers were concerned that Soviet stocks of biological weapons and Soviet expertise might fall into the hands of national states and independent groups hostile to Western interests. The 'microbial threats' might potentially be militarized as 'microbial hostiles'. One indication of the pre-9/11 concern with the intentional release of pathogens was the 1997 speech of Dr David Henderson to the First International Conference on Emerging and Infectious Diseases held in 1998, where he spoke publicly for the first time on bioterrorism.[54] As the former director of the global smallpox eradication campaign at the World Health Organization, Henderson placed his enormous prestige in the service of defining bioterrorism as a public health issue.

After the destruction of the World Trade Center and the subsequent anthrax outbreak the binding together of infectious disease surveillance with military intelligence proceeded with a new level of intensity. In the US, public health became part of the national policy agenda at the new, post-9/11 Department of Homeland Security.[55] In December 2001 the US Congress allocated U$918,000,000 to state and local public health authorities through the CDC to improve their capacity to respond to bioterrorist attacks and public health emergencies.[56] The Public Health Security and Bioterrorism Preparedness and Response Act (2002) was passed with the intent of preventing the use of biological weapons on US soil, and bolstering response capacity in the event of their deployment. The development of new techniques for infectious disease surveillance systems, including 'syndromic surveillance', has been funded through these post-9/11 opportunities. Syndromic surveillance was being developed in the late 1990s,[57] but has flourished under post-9/11 financing. However, as Fidler discusses in chapter 11, the focus since 2001 on emerging infectious diseases and bioterrorism in the CDC's infectious disease control research and programming[58] together with the attendant monies flowing into bioterrorism preparedness in the US public health system, have both been criticized for competing with and undermining core public health practices throughout the United States.[59]

Like GPHIN, syndromic surveillance is an online early warning outbreak detection technique. In the United States, syndromic surveillance has been richly funded as a technique that might provide early warning of bioterrorist events. The techniques of syndromic surveillance have, like GPHIN, been driven by the pressure for 'timeliness', that is, the earliest possible outbreak alert. The demand for 'timeliness' has led to the fashioning of online techniques that *push detection prior to diagnosis*. This is the case with the new 'syndromic surveillance' techniques, which attempt to recognize outbreaks by identifying unanticipated clusters of data. In public health scholarship, the technique of syndromic surveillance contrasts with what is increasingly called 'traditional' infectious disease surveillance based on case reports of suspected or diagnosed cases of specific diseases that are brought to the attention of public health authorities.[60] One 'practical guide' for setting up syndromic surveillance systems makes clear the breadth of data sources that may be used:

> The principal underlying premise of these systems is that the first signs of covert biological warfare attack will be clusters of victims who change their behaviour because they begin to become symptomatic... When people become sick, they may make purchases such as facial tissues, orange juice, and over-the-counter remedies for colds, asthma, allergies, intestinal upsets and so on. They may not report to school or work. Less traditional data sources include work and school absenteeism and retail sales of groceries and over-the-counter medication, including electrolyte products for pediatric gastroenteritis. The next level of detectable activity is likely to be encounters with the health care system. Patients may phone in to nurses or physicians. They may visit sites of primary care, activate 911 emergency medical services, visit emergency departments or be hospitalized. They may have laboratory tests ordered. Some may die. All of this activity may precede the first confirmed diagnosis of a bioterrorism victim.[61]

Now buying facial tissues and skipping school are social practices with many potential meanings. Within public health epidemiology, syndromic surveillance has been criticized as intrinsically flawed by a lack of specificity, that is, too many false positives, too many false alarms.[62]

GPHIN and other early warning online outbreak detection techniques such as syndromic surveillance can only exist under conditions where electronic data is widely available. Such electronic data streams characterize the Global North rather than the Global South, and thus syndromic surveillance cannot at this point in time or for the foreseeable future be used for global surveillance. Unlike GPHIN, syndromic surveillance is organized around *national* needs, defense needs. Again, unlike GPHIN, syndromic surveillance is quantitative rather than based on the analysis of news sources. Global public health surveillance cannot be based on case reports because national surveillance systems are weak outside the richest countries of this world; it is unlikely that large parts of Africa, Asia and South America will have online infectious disease data in the foreseeable future, nor is it desirable in public health terms that this expenditure should be prioritized. Under these conditions, global online news substitutes for 'traditional' surveillance. It is GPHIN rather syndromic surveillance that has weakened national control over the announcement of outbreak, constructing a global knowledge articulated to national sovereignty.

Canada has not escaped the securitization of public health that has characterized the Global North post-9/11, with the fear of bioterrorism being keyed into the rationale for renewal of the national public health system.[63] The securitization of public health, focused as it has been on infectious disease surveillance, has predictably implicated GPHIN, which is institutionally located in the Division of Counter Terrorism Coordination and Health Information Network at the Centre for Emergency Preparedness and Response (CEPR), in the Public Health Agency of Canada.

GPHIN's potential for intelligence uses first became evident during the SARS outbreak in Toronto during 2003, when GPHIN drew national and international publicity as an information source. During the outbreak, GPHIN supplied updates on the numbers of SARS cases internationally to each daily meeting at the Centre for Emergency Preparedness and Response. As the outbreak continued, GPHIN was asked by the Minister of Health to provide information about airport measures and visa restrictions affecting Canadian passengers traveling internationally. The SARS outbreak raised GPHIN's profile within the Canadian government, and since 2003 GPHIN has responded to requests by Ministers for updates on a variety of topics, for instance avian flu. During the SARS outbreak, the multifunctionality of GPHIN's electronic data became evident to GPHIN and to government agencies in Canada and worldwide. GPHIN's data could be processed in multiple ways to serve the goals of differing users.

The SARS outbreak drew the attention of intelligence agencies to GPHIN, an interest that created a conflict of interest between GPHIN and the WHO, its collaborating partner, as one of our research participants made clear:

> Now a lot of intelligence communities got interested as well and that created a bit of a dilemma because WHO said, 'We cannot be associated with intelligence communities'.
> EM and LW: Why is that?
> Political. Because WHO services all countries equally. And so they can't afford to be associated with any particular intelligence group in any way that would compromise their ability to another country that might be at odds with the West or maybe not. You know Cuba's a member of the WHO, for example. So the armed forces medical intelligence committee wanted to subscribe. That's the intelligence in the United States and WHO was very adamant that they could not be associated with that... They said, 'We can't partner with an organization that might also be serving the needs of the intelligence community'. So we said, 'OK. You're not a partner, you're a client'.[64]

As an international agency, the WHO could not compromise its neutrality by aligning with, or being perceived to align with, the spying agencies and defense needs of its individual member states. The legal and diplomatic needs of the WHO drew a boundary between it and the civilian and military intelligence of nations in the Global North. GPHIN, however, wished to develop its capacity for intelligence, doing so by revising its agreement with the WHO, transforming the WHO from a partner to a 'client'.

GPHIN's ongoing funding needs undermined its partnership with the WHO. In order to cover its operational budget, GPHIN moved to a cost recovery system in September 2004, charging its users a subscription fee, and tailoring its online services to meet client needs. The research and development costs for state-of-the-art software capable of functioning in all six languages of the WHO – a second phase of GPHIN's development – were in part funded through a collaborative research agreement with the private sector and a US foundation, the Nuclear Threat Initiative. Consistent government underfunding of GPHIN has required it to initiate innovative public-private sector collaborations and creative approaches to the meaning of 'procurement' between the ministries of differing nations. GPHIN is often touted internationally as a great Canadian contribution to global public health, but its funding needs have met with minimal domestic support.[65]

The situation with respect to GPHIN under-funding has led to it sell its services to organizations that were in a position to pay. At the time of our field research in July 2004, GPHIN had close to one hundred users, mainly health and defense ministries. Military intelligence, including the US

Armed Forces, has access to GPHIN, which has developed the capacity to develop 'warning area alerts' directed to intelligence uses for its clients:

> More recently we are able to provide warning area alerts... For example, Honduras has recently reported several times that they have found training camps within Honduras that are Al-Qaeda supported so this type of information we would provide to our military intelligence officer, Dr. ___, and to other users of GPHIN who would find this information valuable.[66]

The results of chemical spills, which might be either intentional or non-intentional, would be communicated to CSIS (the Canadian intelligence agency), or outbreaks in areas where Canadian troops were stationed overseas would be communicated to the Ministry of Defense.

The sophisticated automated translation capacity found in the new GPHIN platform that was activated in September 2004 gave GPHIN new capacities with potential intelligence uses that would make it attractive to clients with sufficient budgets. One possible area of future intelligence interest in GPHIN involves its capacity to track bioterrorist outbreaks in countries of the Global South that act as transportation hubs for traffic into the Global North. The weak national surveillance systems in the Global South would make these regions vulnerable to bioterrorist events, whereas such events would be quickly identified by the national surveillance capacities of the Global North.[67] Thus, GPHIN's multilingual capacity would provide early warning detection of bioterrorist events that could spread to the Global North via airline travel from the Global South, a feature that would make GPHIN attractive for the biosecurity purposes of the Global North.

The tactics of securing the Global North against bioterrorism have given new meaning to health protection. Where health protection is formulated as a matter of security against terrorism, civilian public health becomes informed by the relevances of defense policy, coordinated with twenty-first century intelligence and the military. One might wonder how this will affect the institutional form, priorities, personnel and actions of global public health with respect to future infectious disease control. At issue here are the tactical valences of global public health surveillance in the twenty-first century, specifically whether it will serve the growth of global democracy and peace in the twenty-first century or the defense and military needs of the Global North. The stakes ultimately involve the politico-ethical form of global biopolitics, specifically whether it shall be run by an elite minority in the Global North in an oligarchic fashion or whether it will serve the health and democratic needs of the people globally.

Conclusion

In this chapter we have described and analyzed the formation of global public health surveillance through a case study of the Global Public Health Intelligence Network. Contemporary global public health surveillance differs from the international knowledge of infectious disease during the period 1945–1990 with respect to its organization of time, social action and geopolitical borders. GPHIN is a technique that enables timeliness of report, that is, a speeding up of international reports of outbreak to a point when outbreak occurs and intervention is possible. GPHIN and other forms of online early warning outbreak detection have created a truly global knowledge of outbreak concurrent with the time of outbreak. Because global public health surveillance operates in that specialized temporal interval known as 'real time', it makes possible an expansion in the volume and institutional significance of previously existing forms of WHO action: official country inquiries and field response. With respect to geopolitical borders, GPHIN created a source of knowledge that was an alternative to official country notification of outbreak. Online early warning outbreak detection not only bypasses national control, it permits the WHO to have an expanded knowledge of many kinds of outbreak in excess of the reportable diseases that required compulsory country notification in the early 1990s: cholera, plague, and yellow fever. In this sense, GPHIN has been vital to the revision of the International Health Regulations which have expanded the scope of international disease reporting to include all public health emergencies considered to be of international concern.[68] The WHO thus has a source of knowledge beyond national control that it uses to act as a lever to pressure countries to divulge information about outbreak.

Global public health surveillance takes apart national secrecy about outbreak and epidemic as well as national ignorance of these events. However, it must not be imagined that global public health proceeds without reference to the space of the nation and national health systems. Rather, global public health surveillance is a level of governmental knowledge that only becomes authoritative and actionable when it is articulated to national verification as well as national consent to the entry of response teams organized through the WHO. The global space of public health surveillance operates *through* national health systems. What we see in the case study of GPHIN is an example of the increasing density of global public health in terms of its knowledge forms and institutional action.

The governmental techniques of early outbreak detection and verification together with global field response are distinct form older, colonial forms of 'containment' through *cordons sanitaires*. Our contemporary concern is less to contain pre-existing epidemics than to identify and directly intervene in local outbreaks regarded as potential or actual international health emergencies. This is not simply a new function for international health response capacity. At its normative best, global public health surveillance aims to protect global population rather than national or colonial ones. It is a new spatial organization of the geopolitical borders of infectious disease control. Global public health surveillance is not organized around a concept of an external frontier to be defended. It is 'empire' without an outside rapidly being integrated into the intelligence needs of the Global North.[69]

Notes

This text is produced with the equal contribution of both authors. We alternate first author position in our joint publications for the purpose of equally distributing public perceptions of primary authorship.

1 D.L. Heymann and G.R. Rodier, 'Hot Spots in a Wired World: WHO Surveillance of Emerging and Re-emerging Infectious Diseases', *The Lancet Infectious Diseases*, 1 (2001): 345–535; T.W. Grein et al., 'Rumours of Disease in the Global Village: Outbreak Verification', *Emerging Infectious Diseases*, 6 (2000): 97–102; D.P. Fidler, 'Emerging Trends in International Law Concerning Global Infectious Disease Control', *Emerging Infectious Diseases*, 9 (2003): 285–90; B. Fantini, 'International Health Organizations and the Emergence of New Infectious Diseases', *History and Philosophy of the Life Sciences*, 15 (1993): 435–57.

2 D.P. Fidler, 'Germs, Governance, and Global Public Health in the Wake of SARS', *Journal of Clinical Investigation*, 113 (2004): 799–804.

3 K. Loughlin and V. Berridge, *Global Health Governance: Historical Dimensions of Global Governance* (London and Geneva: London School of Hygiene & Tropical Medicine and World Health Organisation, 2002).

4 The Conferences were established in response to European anxiety about infectious diseases from the East, especially cholera and yellow fever. Authorities were concerned about the effectiveness of established quarantine measures to deal with the increased movement of goods and people across national borders that had arisen as part of increased international trade. For a discussion of the conferences see N.M. Goodman, *International Health Organizations and their Work* (Edinburgh: Churchill Livingstone, 1971).

5 K. Lee (ed.), *Health Impacts of Globalization: Towards Global Governance* (London: Palgrave, 2003); M. Reich (ed.), *Public-Private Partnerships for Public Health* (Cambridge, Mass.: Harvard Center for Population and Development Studies, Distributed by Harvard UP, 2002); K. Buse and G. Walt, 'The World Health Organisation and Global Public-Private Health Partnerships: In Search of 'Good' Global Governance', in Reich (ed.), *Public-Private Partnerships for Public Health*, pp. 169–98; K. Lee, K. Buse and S. Fustukian (eds), *Health Policy in a Globalising World* (Cambridge: Cambridge University Press, 2002); R. Dodgson, K. Lee and N. Drager, *Global Health Governance: A Conceptual Review* (London and Geneva: London School of Hygiene & Tropical Medicine and World Health Organization, 2002); I. Kickbusch, 'The Development of International Health Policies – Accountability Intact?', *Social Science & Medicine*, 51 (2000): 383–9.

6 Eric Mykhalovskiy and Lorna Weir, 'The Global Public Health Intelligence Network and Early Warning Outbreak Detection: A Canadian Contribution to Global Health', *Canadian Journal of Public Health* 97 (2006): 42–4. *Special issue on Global Health.*

7 On 26–27 July 2004 we interviewed 10 staff and managers of the Global Public Health Intelligence Network (GPHIN). The interviews took place at the Centre for Emergency Preparedness (Health Canada) in Ottawa, Canada. In order to protect the personal confidentiality of the research participants, we have chosen to identify interviewees solely as 'GPHIN Interview, 26–27 July 2004'. All interviews were taped and later transcribed.

8 David Fidler was the first to use the concept of Westphalian and post-Westphalian surveillance of infectious disease. See D. Fidler, 'SARS: Political Pathology of the First Post-Westphalian Pathogen', *Journal of Law, Medicine and Ehtics*, 31 (2003): 485–505.

9 See for example the definition of public health surveillance in the Report of the CDC (Centers for Disease Control and Prevention – Bethesda, Maryland) Guidelines Working Group, *Updated Guidelines for Evaluating Public Health Surveillance Systems* (2001), p. 2, where it is construed as '...the ongoing, systematic collection, analysis, interpretation, and dissemination of data regarding a health-related event for use in public health action to reduce morbidity and mortality to improve health'.

10 Lenore Manderson has done groundbreaking work documenting international reports of outbreak at the Eastern Bureau of the League of Nations Health Organization from 1925 to 1942. The areas of East Asia and Australia were connected via telegraph and radio. See L. Manderson, 'Wireless Wars in the Eastern Arena: Epidemiological Surveillance, Disease Prevention and the Work of the Eastern Bureau of the League of National Health Organisation, 1925–1942' in P. Weindling (ed.) *International Health Organizations and Movements, 1918–1939* (Cambridge: Cambridge University Press, 1995).

11 A.D. Langmuir, 'The Surveillance of Communicable Diseases of National Importance', *The New England Journal of Medicine* 268 (1963): 182–92 and 'William Farr: Founder of Modern Concepts of Surveillance', *International Journal of Epidemiology* 5 (1976): 13; S.B. Thacker, 'Historical Development' in S. Teutsch and R.E. Elliott (eds), *Principles and Practice of Public Health Surveillance* (Oxford: Oxford University Press, 2000) p. 11; S.B. Thacker and R.L. Berkelman, 'Public Health Surveillance in the United States', *Journal of the American Medical Association*, 249 (1988): 166.

12 S.B. Thacker and M.B. Gregg, 'Implementing the Concepts of William Farr: The Contributions of Alexander D. Langmuir to Public Health Surveillance and Communications', *American Journal of Epidemiology*, 144 (8 Suppl) (1996): S23.

13 *Ibid.*, p. S26.

14 For a discussion of the 1973 smallpox outbreak and its impact on disease surveillance in Britain, see G. Pollock, *Fevers and Cultures: Lessons for Surveillance, Prevention and Control* (Abingdon: Radcliffe Medical Press, 2003) p. 27.

15 O. Aginam, 'International Law and Communicable Diseases', *Bulletin of the World Health Organisation*, 80 (2002): 946–50; J. Fricker, 'Emerging Infectious Diseases: A Global Problem', *Medicine Today*, 6 (2000): 334–445; D.P. Fidler, 'Public Health and National Security in the Global Age: Bioterrorism, Pathogenic Microbes and *Realpolitik', George Washington International Law Review*, 25 (2003): 787–856; Heymann and Rodier, 'Global Surveillance of Communicable Diseases', *Emerging Infectious Diseases*, 4 (1998): 362–65; Heymann and Rodier, 'Hot Spots in a Wired World'.

16 Heyman and Rodier, 'Global Surveillance of Communicable Diseases'.

17 J. Lederberg, R.E. Shope and S.C. Oaks Jr. (eds), *Emerging Infections: Microbial Threats to Health in the United States* (Washington, DC: National Academy Press, 1992).

18 *Ibid.*, p. 34.
19 Centers for Disease Control and Prevention, *Addressing Emerging Infectious Disease Threats: A Prevention Strategy for the United States* (1994), http://www.cdc.gov/ncidod/publications/eid_plan/Default.htm. (accessed 4 May 2006)
20 L. Garrett, *The Coming Plague: Newly Emerging Diseases in a World Out of Balance* (New York: Farrar, Straus and Giroux, 1994), pp. 604–651.
21 World Health Organisation, *Global Outbreak Alert and Response: Report of a WHO Meeting* Department of Communicable Diseases Surveillance and Response (Geneva WHO, 2000) http://who/cds/csr/2000.3, pp. 7–9 (accessed 4 May 2006)
22 J.W. Leduc, 'Action Plan of the World Health Organisation', *Global Issues: An Electronic Journal of the United States Information Agency* 1, 17 (Nov. 1996), http://usinfo.state.gov/journals/itgic/1196/ijge/gj-7.htm. (accessed 4 May 2006).
23 'WHO at Fifty. 4 Highlights of Activities from 1989 to 1998', *World Health Forum* 19 (1998): 452.
24 D.L. Heymann and G.R. Rodier, 'Global Surveillance, National Surveillance, and SARS', *Emerging Infectious Diseases*, 10 (2004): 173–5.
25 Global Outbreak and Response Network, Webpage, http://www.who.int/csr/outbreaknetwork/en/ (accessed 4 May 2006).
26 Heymann and Rodier, 'Global Surveillance'.
27 Interestingly, similar arguments about the effectiveness of quarantine and reduced travel times associated with the introduction of steam ships surfaced during the deliberations of the international sanitary conferences (see Loughlin and Berridge, *Global Health Governance*).
28 For our purposes, a Westphalian system of world health existed from the period after World War II until the mid-to-late 1990s, that is, from the dissolution of colonial empires to the invention of global public health surveillance and response.
29 M. Hardiman, 'The Revised International Health Regulations: A Framework for Global Health Security', *International Journal of Antimicrobial Agents*, 21 (2003): 207–11.
30 GPHIN Interview, 26–27 July 2004.
31 For example, it is estimated that the 1991 cholera outbreak in South America cost Peru approximately US$150 million in lost tourism and $750 million in lost trade. Estimates of the cost of the 1994 bubonic plague in India are in the range of US$2 billion. See R.A. Cash and V. Narasimhan, 'Impediments to Global Surveillance of Infectious Diseases: Economic and Social Consequences of Open Reporting', *Bulletin of the World Health Organisation*, 78 (2000): 1358–67.
32 M. Medina, 'Time Management and CNN Strategies (1980–2000)' in A.B. Albarran and A. Arrese (eds), *Time and Media Markets* (Mahwah, New Jersey: Lawrence Erlbaum, 2003) pp. 81–95.
33 GPHIN Interview, 26–27 July 2004.
34 M.W. Zacher, 'Global Epidemiological Surveillance: International Cooperation to Monitor Infectious Diseases' in I. Kaul, I. Grunberg and M. Stern (eds), *Global Public Goods: International Cooperation in the 21st Century* (Oxford: Oxford University Press, 1999) pp. 266–83.
35 Cash and Narasimhan, 'Impediments to Global Surveillance of Infectious Diseases'.
36 GPHIN Interview, 26–27 July 2004.
37 GPHIN Interview, 26–27 July 2004.

38 Heymann and Rodier, 'Global Surveillance of Communicable Diseases'; D.L. *Heymann, Food Safety, an Essential Public Health Priority* (2002), http://www.foodsafetyforum.org/global/opening3_en.htm. (accessed 4 May 2006).
39 J.P. Woodall, 'Global Surveillance of Emerging Diseases: The ProMED-mail experience', *Cad Saude Publica*, 17 (Suppl.) (2001): 147–54; M. Hugh-Jones, 'Global Awareness of Disease Outbreaks: The Experience of ProMED-mail', *Public Health Reports*, 116 (Supplement 2) (2001): 27–31.
40 GPHIN interview, September 26, 2005 and R. St. John, GPHIN Presentation (2004), http://www.dtra.mil/ASCO/wpc/session3/St%20John.%20GPHIN%20Presentation_%20Jan%2020%202004.pdf. (accessed 4 May 2006).
41 GPHIN Interviews, 26–27 July 2004.
42 L. Garrett, 'Understanding Media's Response to Epidemics', *Public Health Reports*, 116 (Supplement 2) (2001): 88.
43 GPHIN Interview, 26–27 July 2004.
44 David Henderson, 'Surveillance of Smallpox', *International Journal of Epidemiology*, 5 (1976): 19–28.
45 GPHIN Interview, 26–27 July 2004.
46 It is a misperception to read the history of public health surveillance as based in its entirety on case-reports. News has been a component of public health surveillance since the 1955 polio outbreak when CDC epidemiologists used newspaper reports as part of their data (A. Langmuir, 'The Surveillance of Communicable Diseases of National Importance', *New England Medical Journal*, 268 (1963): 182–92). During the 1980s, public health epidemiology began to draw on telephone surveys (S.B. Thacker, R.L. Berkelman and D.F. Stroup, 'The Science of Public Health Surveillance' *Journal of Public Health Policy* 10 (1989): 190.
47 GPHIN Interview, 26–27 July 2004.
48 'China and the Internet', *The Economist* (27 April 2006); Andy Ho, 'China Bug – Is It Ebola-like Bird Flu?' *Yale Global Online* http://yaleglobal.yale.edu/display.article?id=6077 (accessed 2 May 2006).
49 Centers for Disease Control and Prevention, 'Update on Emerging Infections: Syndromic Surveillance for Bioterrorism Following the Attacks on the World Trade Centre – New York City, 2001', *Annals of Emergency Medicine*, 41 (2003): 414–18; A. Reingold, 'If Syndromic Surveillance is the Answer, What is the Question?', *Biosecurity and Bioterrorism: Biodefense Strategy, Practice and Science*, 1, 2 (2003): 77–81.
50 Lederberg et al., 'Emerging Infections: Microbial Threats to Health in the United States'.
51 Centers for Disease Control and Prevention, Addressing *Emerging Infectious Disease Threats: A Prevention Strategy for the United States.*
52 ProMED-Mail Interview, 27 July 2005.
53 L. Garrett, *Betrayal of Trust: The Collapse of Global Public Health* (New York: Hyperion, 2000), pp. 486–550.
54 *Ibid.*, p. 450.
55 I. Kickbusch, 'The Contribution of the World Health Organization to a New Public Health and Health Promotion', *American Journal of Public Health*, 93 (2003): 383–9.
56 S.W. Marmagas, L.R. King and M.G. Chuk, 'Public Health's Response to a Changed World: September 11, Biological Terrorism, and the Development of an Environmental Health Tracking Network', *American Journal of Public Health*, 93 (2003): 1228.

57 J. Koplan, 'CDC's Strategic Plan for Bioterrorism Preparedness and Response', Public Health Reports, 116 (Supplement 2) (2001): 13; D.M. Sosin, 'Syndromic Surveillance: The Case for Skilful Investment', *Biosecurity and Bioterrorism: Biodefense Strategy, Practice and Science*, 1 (2003): 247.
58 J. Koplan, 'CDC's Strategic Plan for Bioterrorism Preparedness and Response', p. 9
59 Reingold, 'If Syndromic Surveillance Is the Answer, What Is the Question?'
60 M.S. Green and Z. Kaufman, 'Surveillance for Early Detection and Monitoring of Infectious Disease Outbreaks Associated with Bioterrorism', *Israel Medical Association Journal*, 4, 7 (2002): 503; Sosin, 'Syndromic Surveillance: The Case for Skilful Investment', p. 247.
61 K.D. Mandl et al., 'Implementing Syndromic Surveillance: A Practical Guide Informed by the Early Experience', *Journal of the American Medical Informatics Association*, 21 (November 2003): 11.
62 Reingold, op. cit.; M.A. Stoto, M. Schonlau and L.T. Mariano, 'Syndromic Surveillance: Is It Worth the Effort?', *Chance*, 17, 1 (2004): 19–24.
63 National Advisory Committee on SARS and Public Health, *Learning from SARS: Renewal of Public Health in Canada* (Ottawa: National Advisory Committee on SARS and Public Health, 2003), p. 2
64 GPHIN Interview, 26–27 July 2004.
65 Mykhalovskiy and Weir 'The Global Public Health Intelligence Network and Early Warning Outbreak Detection', pp. 42–4.
66 GPHIN Interview, 26–27 July 2004.
67 GPHIN Interview, 26–27 July 2004.
68 M. Hardiman 'The Revised International Health Regulations: A Framework for Global Health Security'. *International Journal of Antimicrobial Agents*, 21 (2003): 207–10.
69 On 'empire without an outside' see M. Hardt and A. Negri, *Empire* (Cambridge, Mass.: Harvard University Press, 2000).

Index